T0161776

POWERING A CITY

How Energy and Big Dreams Transformed San Antonio

Catherine Nixon Cooke

Foreword by Bill Greehey
Introduction by Char Miller

Maverick Books / Trinity University Press
San Antonio, Texas

Published by Maverick Books, an imprint of Trinity University Press
San Antonio, Texas 78212

Book design by Seale Studios

Cover image courtesy of CPS Energy

ISBN 978-1-59534-843-2 paperback
ISBN 978-1-59534-857-9 hardcover
ISBN 978-1-59534-844-9 ebook

Trinity University Press strives to produce its books using methods and materials in an environmentally sensitive manner. We favor working with manufacturers that practice sustainable management of all natural resources, produce paper using recycled stock, and manage forests with the best possible practices for people, biodiversity, and sustainability. The press is a member of the Green Press Initiative, a nonprofit program dedicated to supporting publishers in their efforts to reduce their impacts on endangered forests, climate change, and forest-dependent communities.

The paper used in this publication meets the minimum requirements of the American National Standard for Information Sciences—Permanence of Paper for Printed Library Materials, ansi 39.48–1992.

CIP data on file at the Library of Congress

21 20 19 18 17 | 5 4 3 2 1

For the more than 10,000 men and women
who have worked for CPS Energy and its predecessors,
powering San Antonio and its dreams since 1860

At the heart of putting **People First** is CPS Energy's commitment to caring for our customers and community. With your purchase of this book, you help neighbors in need to pay their energy bills through the Residential Energy Assistance Partnership. This 501(c)3 exists to provide financial assistance for energy services to qualifying households with young children, or with medically dependent or elderly family members.

FOREWORD

Bill Greehey

Most San Antonians probably do not know how truly blessed we are to have the nation's largest municipally owned gas and electric utility serving our hometown. I think most of us are aware that CPS Energy provides thousands of well-paying jobs with great benefits for San Antonio employees and their families. And in times of crisis, such as the treacherous tornadoes that mangled power poles into toothpicks in 2017, we see these brave men and women work tirelessly around the clock to get our power running again.

When we pay our utility bills, I doubt many of us consider the forward-thinking leaders of the 1970s and 1980s who led the effort to develop the South Texas Project and open our first coal-fired power plant in the midst of the energy crisis. Most of us probably don't realize that these innovations have made it possible for us to enjoy some of the lowest energy costs among the nation's twenty largest cities. And I sincerely doubt many know that 14 percent of CPS Energy's revenues are remitted to the City of San Antonio's general fund each year, accounting for 20 percent of the city's annual operating budget.

In recent years CPS has continued its innovation, first with the J. K. Spruce coal plant built in the 1990s, featuring more than $100 million in environmental protection systems, and again in 2010 with the J. K. Spruce 2 plant, one of the nation's cleanest coal units, with more than $250 million in the best emissions control equipment. Today CPS Energy is leading the way in energy capacity among municipally owned utilities across the United States with a goal of generating 20 percent of its power from renewable sources by 2020, including 400 megawatts from solar power.

For the past seventy-five years CPS has been among our community's greatest contributors. Its seventy-fifth anniversary in 2017 is a good time to reflect on the ways CPS has helped shape our city and improve our quality of life. One of these ways may surprise you—the little-known story of how Valero Energy Corporation was born out of the ashes of a tremendous civic firestorm involving natural gas shortages, communitywide blackouts, and billions of dollars in lawsuits. Although my early (and fiery) encounters with the City of San Antonio and CPS were difficult, a partnership was eventually forged that resulted in decades of success on many levels for our community.

It all started in May 1973, when I served as senior vice president of finance for Coastal States Gas Producing Company, the parent company of LoVaca Gathering Company. LoVaca was the largest intrastate natural gas pipeline operator in Texas, and the company had contracts to directly or indirectly supply natural gas to most of the state. The City of San Antonio, through City Public Service, was LoVaca's largest customer. Those contracts had LoVaca locked into selling gas for about twenty cents per cubic foot. When gas came into short supply in the 1970s, gas prices skyrocketed to more than two dollars per thousand cubic feet, and LoVaca was unable to honor the contracts. Customers sued for billions of dollars.

San Antonio was hit especially hard with gas curtailments. There were blackouts. The lights weren't on at the airport, and no one could have Christmas lights. It was a real state of crisis, and people were upset and, in some cases, hostile. LoVaca asked the Texas Railroad Commission— the state agency responsible for natural gas—for rate relief, realizing that they would go bankrupt without it. Coastal States and LoVaca entered into a judgment to turn over the management of LoVaca to the 500th District Court judge in Austin. He appointed an independent, five-person board to run LoVaca, and I was appointed by Coastal States to monitor the board meetings.

But LoVaca continued to lose money because the rate relief was not sufficient to make a profit. The judge fired two of the appointed board members, and then the board fired the acting president. These were chaotic times, to say the least! In April 1974 the board hired me to serve as president and CEO of LoVaca. I resigned from Coastal States and reported directly to the LoVaca board. My first priority was to meet with all of the major customers to establish credibility for the new board. Since CPS was our largest customer, I met with them first.

The first meeting included CPS General Manager Tom Deely, CPS board chair Tom Berg, and board member Glenn Biggs. They were in favor of a settlement to ensure that San Antonio would maintain a stable supply of natural gas. Tom Berg, in particular, was publicly attacked for his position, since most customers wanted to proceed with the lawsuit. Despite the pressure, CPS leadership knew that a settlement was in San Antonio's best long-term interests.

I then met with H. B. "Pat" Zachry, one of the city's most highly respected business leaders. He outlined the basis of a settlement. Pat told me that Oscar Wyatt, the notorious founder and CEO of Coastal States, whom customers blamed for defaulting under the contracts, "would have to bleed." He meant that LoVaca should be spun off from Coastal States so that it would always be managed independently from Oscar. Pat suggested that Oscar, who was the largest owner of Coastal States stock, should receive no stock in the spin-off; instead, customers should own a large share of the stock with the requirement that it be sold over a period of time. Pat also suggested that Coastal States would have to "bleed" as well, and his plan envisioned the company spending hundreds of millions of dollars drilling for natural gas and selling the gas at discounted prices to customers.

I started working with all of our customers and the courts on a settlement that would result in LoVaca being permanently owned and managed separately from Coastal States. Despite the support I had from CPS and business and civic leaders in San Antonio, there was still major opposition from many customers, and most people felt that a settlement was impossible. Fortunately Mayor Lila Cockrell called for a public hearing, which helped turn public sentiment in favor of a settlement. The negotiations took more than six years, but we finally succeeded. The final settlement mirrored the plan Pat Zachry had originally outlined—with one important addition that involved a major risk to LoVaca, and to my career.

To keep CPS and San Antonio committed to the settlement, we made the decision to move our corporate headquarters from Houston to San Antonio. We constructed a headquarters building, hired several hundred people for the new company, and built a computer facility—a year and a half before the settlement was completed! (Ironically, those buildings will soon become the corporate headquarters for CPS Energy.)

This was a huge risk because if we had not reached a settlement, we would have been sued for hundreds of millions of dollars. The move to San Antonio helped us seal the deal, however, and on January 1, 1980, we agreed to a $1.6 billion settlement, at the time the largest corporate spin-off in U.S. history. The company, now fully independent from Coastal States, became Valero Energy Corporation.

Throughout the settlement negotiations I made a pledge to San Antonio that Valero would be a great corporate citizen, and I'm proud that we more than lived up to that pledge as Valero continued to grow and prosper. It became one of San Antonio's largest employers and one of the best companies to work for, eventually ranking third on *Fortune* magazine's list of the 100 Best Companies to Work For. We took pride in our caring and sharing spirit, with our company and employees giving generously to virtually every charity and civic cause in the community. Our employees also gave their time and talent, contributing tens of thousands of volunteer hours a year to a wide range of charitable and civic causes.

Then, in 2007, NuStar Energy spun off from Valero, and it has built on the successful corporate culture that we started. NuStar continues to be recognized among the best employers in America and is known for having one of the city's strongest charitable giving and volunteer programs.

Not only did the vision, support, and cooperation of the leadership at CPS and the City of San Antonio help make the settlement with LoVaca possible, ensuring that citizens of San Antonio were able to maintain a long-term, reliable source of energy at very competitive rates, but they also helped pave the way for two outstanding corporate citizens to become part of our community. And, of course, we are all fortunate that CPS Energy has established itself as an innovative company that provides outstanding service and value to its customers, great jobs with good pay and benefits to its more than three thousand employees, and tremendous support to countless causes throughout the community.

What a tremendous asset CPS Energy is—and has been—to our community. I am proud of their many contributions to San Antonio over the past seventy-five years, and I know they will be just as successful in the next seventy-five.

INTRODUCTION

Char Miller

Resources make a city. The flows of energy, water, and waste, like land-use patterns and mobility, help define the urban economy, political organization, and class structures. Put in different terms, how a community identifies, develops, manages, and allocates resources determines how sustainable, resilient, and just it is (or is not).

Exploring some of these intertwined issues is *Powering a City*, Catherine Nixon Cooke's compelling and beautifully illustrated history of CPS Energy. The City of San Antonio owns the utility, which provides the nation's seventh largest city with the electricity and natural gas it consumes. As vital as these resources are to the community's economic fortunes and its standard of living—the two being parts of a whole—CPS Energy also contributes a stream of capital into the city's budget, returning 14 percent of its gross revenues to San Antonio. These funds make up roughly 20 percent of the Alamo City's annual operating budget. Cooke reports that the utility has infused more than $7 billion into the city's coffers since its founding in 1942. For the past seventy-five years, power and money have proved an unbeatable combination.

This winning arrangement resulted from the city doing something unusual amid the carnage of World War II. It purchased the San Antonio Public Service Company. The acquisition, which the Public Utility Holding Company Act of 1935 enabled, did exactly what its advocates, President Franklin Roosevelt and the New Dealers, had desired. City leaders around the country used the new law's regulatory authority to break up the large-scale monopolies that had dominated energy-resource management. SAPSCo, for example, was a unit of the multistate holding company United Light & Power, which in turn American Light & Traction owned; this latter entity was under the control of Emerson McMillin, a Wall Street banking house. The details matter. It was a progressive piece of federal legislation alone that made it possible for the city to secure control of this privately owned utility. The newly formed City Public Service Board gave San Antonio a significant measure of control over its future and has been paying huge dividends ever since.

Although *Powering a City* opens with this pivotal moment and might have carried the narrative forward to the present—that's what an ordinary commemorative volume might do—Cooke adopts a much more engaging approach by going back in time. In so doing, she contextualizes the history of resource use in the San Antonio River valley, beginning with the arrival of Spanish colonizers in 1718. To build San Antonio, the Spanish had to ensure that this remote outpost in northern New

Spain had access to two essential resources, water and wood. It was not by happenstance that its military officers and Catholic missionaries, when they encountered the environs surrounding San Pedro Creek and the San Antonio River, immediately estimated streamflow and cataloged the variety of trees rooted in the landscape. Their rough calculations dovetailed with the use that the Payaya and their ancestors had made of this well-watered and fertile valley for millennia. Through persuasion, coercion, and conversion, the Spanish controlled native labor to create a settlement that consisted of the missions, a civilian community, and a presidio. The Payaya and other South Texas Indians dug the original acequias that provided water for irrigation and household use; they and the Spanish settler-colonists began to harvest nearby trees and shrubs to provide fuel for cooking and illumination. Not everyone had equal access to these resources. When in 1731 the Canary Islanders arrived as part of the Spanish crown's effort to increase the community's population, they immediately secured control of the acequias' flow and the town's most arable lands. Their dominance, lasting until the late eighteenth century, is a reminder of the degree to which social status and political clout are bound up with manipulation of critical resources.

Succeeding generations of San Antonians have continued to contest with one another over the control and distribution of these resources and new ones that subsequently became available. Since the eighteenth century the community has fought over water, a struggle that became more fraught as entrepreneurs introduced ever more sophisticated waterworks to capture streamflow and pumps, reservoirs, and pipes to distribute it across the community—to those who could afford to buy it. This inequality continued well after the city purchased the local water company. Because the Anglo power elite kept "water out of 'politics' for decades,"

political scientist Heywood Sanders has observed, "the city's leadership created a system that discriminated on the basis of wealth." It was not until the 1970s that the City Water Board—now SAWS—finally began to serve the poorest neighborhoods.

A similar skewing occurred with innovations in transportation. The arrival of the railroad in 1877, and shortly thereafter of an expanding system of streetcars, altered how some people rolled across the growing city, while others walked. These new forms of transit reinforced the widening physical gap between the rich and the rest. Developers constructed streetcar suburbs well beyond the central core, a separation of the haves and have-nots that the automobile and local expressways such as Highway 281, Interstate 10, and Interstate 35 would extend across a sprawling terrain.

The power grid tells the same story. In text and image, Cooke lays out the initial efforts to illuminate San Antonio, starting in 1860 with the wood resin–burning gasworks that lit up some streets and Main Plaza, as well as elite homes gathered around San Pedro Park. Early adopters included, perhaps more garishly, some of the community's most notorious saloons. As gas lines snaked out to those storekeepers, commercial sites, and homeowners who could pay, the increasing demand brought competition and complication. One story captures this well. In 1889 the San Antonio Edison Company secured a franchise to electrify streetcars, and its power plant began to provide electricity to businesses and residences located along the rail lines. Alas, because the operation initially used direct—not alternating—current, every time a trolley rattled past an Edison customer, interior lights dimmed. The San Antonio Gas Company seized this opportunity, purchased its competitor, underwrote the shift to the more stable form of energy, and then had to battle with two other electricity companies for market share. The outcome was predictable, Cooke notes: "All three strung their wires with abandon and began erecting a forest of poles." The resulting visual chaos, beautifully captured in an 1898 photograph of downtown San Antonio, also affirms historian Martin Melosi's observation, in *Effluent America: Cities, Industry, Energy, and the Environment,* that the "energy history of the United States . . . has been an ongoing effort to cope with abundance."

One of those resources that produced so much power—coal—replaced wood as the fuel of choice for power plants across the country and in San Antonio. Another—capital—drove a wave of consolidation in the first decades of the twentieth century as Wall Street financiers snapped up utilities and bundled them in holding companies. The dense web of ownership intentionally obscured the industry's monopolization and the financiers' immense political power in Congress and state capitals. It took progressive governors like Pennsylvania's Gifford Pinchot, who in the mid-1920s exposed the tangled lines of corporate control, and his friend Harold Ickes, an attorney who later served as the secretary of the interior under Franklin Roosevelt, to raise the storm of protest that in 1935 resulted in the Public Utility Holding Company Act. San Antonio's farsighted leaders were not wrong to recognize in the legislation the community's liberation and its potential enlightenment.

Nothing about this envisioned future came easily, and *Powering a City* testifies to just how complicated San Antonio's subsequent development has proved to be. New sources of generative power, from coal and natural gas to nuclear energy, brought with them increased kilowatts and new logistical challenges of transportation, distribution, and pollution. Low-cost energy became a kind of loss leader—a draw for businesses to locate in San Antonio, many of which offered low-wage employment. As transmission towers and power lines followed and accelerated the suburbanization of the urban periphery, San Antonio built over large swaths of the Edwards Aquifer recharge and contributing zones. In the 1970s this complex interplay sparked a powerful protest movement to protect the city's sole source of water, heated debates that have continued to the present. Even CPS Energy's inclusion of renewables in its portfolio, a smart move in wind-rich and sun-drenched Texas, comes with its upsides and down. Like its peer cities along Interstate 10 West—El Paso, Tucson, Phoenix, and Los Angeles—San Antonio moved quickly to capture more sustainable energy sources that produced fewer greenhouse gases. In 2007 this amounted to an impressive 11 percent of its total energy production, a percentage that has increased, but only slightly, over the years. Meanwhile other western megacities have pushed further and faster to wean themselves from fossil fuels and in their escalating commitments to achieving zero emissions, giving them a competitive edge.

How CPS Energy and the City of San Antonio respond to this competition across the rest of the twenty-first century will determine the community's ability to grow wisely and sustainably. Catherine Nixon Cooke, for one, is betting that the publicly owned utility will meet the anticipated challenges, that its past is prologue. "As energy needs grow and change, as fuel sources shift and distribution systems evolve," she concludes, "CPS Energy will continue to adapt. 'People First' remains the company's enduring motto as it powers San Antonio and its dreams for the future."

PREFACE

This is a story about energy and its extraordinary impact on a city. It's also a story about people—some in the public spotlight, some living more ordinary lives—who have seen their dreams for San Antonio, and for their families and future, realized in a myriad of ways.

When I was asked to write the history of CPS Energy, the largest municipally owned utility company in the United States, I worried that it wouldn't be interesting. As the utility company prepared to celebrate seventy-five years of ownership by the City of San Antonio, I wasn't sure that a tale of generators, turbines, power plants, and transmission lines could captivate readers. And although the purpose of the book was to recognize an important milestone in the company's history, I knew the story went much further back—to 1860, when the San Antonio Gas Company began manufacturing gas for streetlights, using tree resin that arrived by oxcart, in a small plant on San Pedro Creek. I wanted to start at the beginning of that story, of course, long before some very smart city leaders purchased a utility from a giant, New York–based holding company in 1942. Luckily CPS Energy was in full agreement, recognizing that its earlier incarnations shaped the company it is today.

I discovered that San Antonio's journey from a dusty, dark frontier town—with more saloons than grocery stores or banks—to the seventh largest city in the United States was fascinating. As I started the research for this book, I realized that the transformation was due largely to power—electricity and gas that lit homes and businesses, ran equipment and machines, provided heat for cooking and warmth, and cooled a city where temperatures can exceed 100 degrees.

Another part of the transformation equation is people, of course, and I was delighted to find all sorts of remarkable characters lurking in the archives at CPS Energy—characters from long ago, captured in historic documents, and current employees who welcomed me into a unique culture and shared their stories.

Over the years the company has maintained a treasure trove of historic photographs—some dating as far back as the early 1900s. Every issue of its in-house magazine, the *Broadcaster*, which began publication in 1922, was bound in beautiful leather binders. Through those pages, I read about the company's first general manager, Col. William Bockhout Tuttle, and his number one priority—the community, which in his mind was comprised of both customers and employees. The discovery of natural gas, the Great Depression and its impact on San Antonio, and World War II, when 25 percent of the company's workforce enlisted in the armed forces, came to life. Tom Shelton, a friend who has worked in the incredible repository of historic photographs at the University of Texas

Institute of Texan Cultures for thirty years, helped me find wonderful illustrations from that collection for this book.

The rise and fall of big holding companies is part of the story as well. Characters like Samuel Insull, who fled to Europe when his empire crashed in 1929, and Secretary of the Interior Harold Ickes, vowing to bring him to justice, add to the story's richness. Once President Franklin Roosevelt's Public Utility Holding Company Act of 1935 became a reality, utility companies like CPS Energy (then San Antonio Public Service Company) were put up for sale by their holding companies—and the race was on to buy what was correctly perceived as a treasure. The competition between the City of San Antonio and the Guadalupe-Blanco River Authority (GBRA), whose champions were U.S. Congressman Lyndon B. Johnson and Texas State Senator Alvin J. Wirtz, was contentious. The outcome on October 24, 1942, resulted in the "deal of the century" for our city. Readers may be surprised to learn that a remarkable trust indenture governs the partnership between the City and the utility company, and that more than $7 billion in CPS revenues have gone into the City's general fund over the years. That money has powered a lot of dreams.

CPS Energy survived the energy crisis of the 1970s, and thanks to a legendary lawsuit over gas contracts with larger-than-life oilman Oscar Wyatt, the city won another real prize when Valero Energy Corporation was established here. That story, too, is riveting and filled with colorful characters.

Tall buildings and big arenas, a sprawling Southwest Research Institute and South Texas Medical Center, HemisFair 1968, SeaWorld, Fiesta Texas, Morgan's Wonderland, and a boom in housing and shopping center construction were—and are—"powered" by CPS Energy. And over the past seventy-five years the company has led the way in fuel diversification and preparedness for the big changes taking place in the energy world. It has developed an impressive presence in renewable energy, especially with its wind and solar projects, and it is ready to meet the future. CEO Paula Gold-Williams describes the company in 2017 as "nimble." It's a wonderful word that captures the innovative, can-do spirit that has been a part of CPS Energy from the beginning.

Social anthropologist Gretchen Bakke noted in her bestseller *The Grid: The Fraying Wires between Americans and Our Energy Future* that most Americans cannot tell the difference between a power plant and an oil refinery. The energy that powers our homes and offices has become an expectation. This anniversary celebration is the perfect time to remember that in some parts of the world, life is not much different than ours was in the 1850s, before light and power catapulted us into modernity. And along with light and power, and substantial revenues, CPS Energy has provided extraordinary community service. Employees pledged more than $1 million to United Way in 2017, and "People First" remains the company's core motto.

My initial worries were unfounded. This story is filled with wonderful characters, plenty of drama, and important information about the world of energy—and I've started looking at transformers and power lines with new appreciation.

CPS Energy sent trucks and crews to Victoria, Texas, to assist with power restoration following Hurricane Harvey in September 2017.

SAN ANTONIO
THE LIGHT
AN INDEPENDENT TEXAS NEWSPAPER
Member of the Associated Press ★ A Constructive Force in the Community

HOME EDITION

VOL. LXII—NO. 278 Published by The Light Publishing Company, San Antonio, Texas. SATURDAY, OCTOBER 24, 1942. FOURTEEN PAGES. THREE CENTS Per copy in the city and vicinity. Five cents on trains and elsewhere.

U. S., British Open Front in Egypt

SAPSCO Purchase Closed by City

Collision of Planes, Crash Fatal to 12

PALM SPRINGS, Calif., Oct. 24. (AP)—After an American airlines official had declared an army bomber was in collision with a transport plane shortly before the latter crashed near here late yesterday, killing 12, a civilian defense airplane spotter told today of seeing the two twin-engined ships come together just before one wavered from its course and struck a mountainside.

Southern California army authorities maintained silence pending a military investigation.

In San Francisco, Fourth air force spokesmen maintained that to their knowledge no military plane was involved.

COLLISION SEEN.

The plane spotter, R. M. Martin, told reporters he saw two planes flying at about 8000 feet, one about a mile and a half behind the other.

"The one in the rear veered away into a cloud and I thought it had changed course.

"Then it came back and slid in so close to the other plane I couldn't distinguish between them. Bits of metal began flying from the planes."

Martin, watching the airliner spin earthward, lost sight of the other ship, and couldn't say what happened to it.

A military guard was thrown about the wreckage.

NINE PASSENGERS.

The crew of three and all nine passengers were killed, airline officials announced. Among the dead was Ralph Rainger, who composed such popular songs as "Love In Bloom," "June In January," and "Here Lies Love."

The statement of the airlines executive, Charles A. Rheinstrom, vice president in charge of traffic, was made in New York and released through the airlines office here. He said:

"American airlines flight 28, eastbound Los Angeles to New York, was in collision with army bomber, crashed half a mile west of Palm Springs at 5:15 p. m. Pacific time Friday, October 23.

"Left Los Angeles at 4:30 p. m. and was at normal cruising altitude on course, clear weather and daylight when accident occurred."

CREW FROM DALLAS.

American Airlines announced the names of the crew members as Capt. Charles F. Pedley, pilot; First Officer L. F. Keppert, co-pilot, and Estelle Regan, stewardess, all of Dallas.

The passenger list also included: B. R. Vest, Birmingham, Ala.; M. C. Henderson, Phoenix, Ariz., member interstate industrial commission; C. Baker, Phoenix; Frank Bird, Lockheed Aircraft corporation employe; L. A. Hege, no address available, E. H. Wallace, Las Vegas, Nev.; Lieut. Joseph R. Rosser, Santa Ana army air base; C. M. Wert, Los Angeles.

Fewer Hobnails Save Metal

LONDON, Oct. 24.—British troops all over the country are quietly celebrating today. They've heard some really good news. This is the announcement issued by the war office that started it all:

"Army boots are to have fewer nails because of the metal shortage. In the future military shoes will have 30, instead of the 38 hobnails."

Axis Planes Raid Libya Allied Post

CAIRO, Oct. 24.—(AP)—Two Axis raids on Kufra oasis, Fighting French outpost in southeastern Libya, were reported today by the Fighting French press service which said little damage was done.

Kufra oasis was captured by De Gaullists in March, 1941. It lies 500 miles south of the Axis-held port of Tobruk.

Details Being Held Up

San Antonio's purchase of the San Antonio Public Service company property was closed at 10 a. m. Saturday.

Mayor C. K. Quin made the announcement, following a conference with interested parties, in Hal Dewar's office.

The city has an operating agreement with the Guadalupe-Blanco authority, which corporation has in turn transferred its agreement to the Lower Colorado authority.

"It is substantially the same deal," said the mayor in dollars and cents," Quin said. He indicated, however, that the new deal might result in some saving to San Antonio.

"Full details of the deal will be made public just as soon as we can possibly do so," said the mayor in announcements to newspapers.

Co-incident with the mayor's announcement came word from Austin that the Guadalupe Electric company of San Antonio had been chartered by the secretary of state.

The capital stock was $1000 and the incorporators are A. E. Robertson, J. D. Williamson and Creston H. Funk.

Reds Crack German Line At Stalingrad

MOSCOW, Oct. 24.—(AP)—Red army troops breached the front line of the Germans' fortified left flank northwest of Stalingrad and the city's garrison repulsed all attacks to hold a northside factory against a tank-supported infantry brigade despite a shower of 1500 Stuka bombs, the Russians said today.

Clouds which recently brought snow evidently had lifted, for the army newspaper Red Star indicated German aerial activity was increased.

2200 NAZIS KILLED

Red Star said the Russian army to the northwest, finding a weakly defended junction between two German units, opened an offensive and broke into the enemy front line.

The Germans counterattacked but, after day long hand-to-hand fighting inside the Nazi positions, they were ejected with a loss of 700 dead and were forced to retreat, Red Star said.

This was the 61st day of siege.

Field reports indicated more than 2200 Germans were killed at Stalingrad yesterday while Soviet ground gunners and fighters were accounting for 27 Nazi planes.

STREAM CROSSED

Loss of a water barrier by Red defenders of the western Caucasus was acknowledged in the noon communique. The invaders, striking southeastward from Novorossisk, were reported to have crossed the stream in the drive upon the port of Tuapse.

The Soviet radio said Soviet warships had attacked an enemy-occupied post on the Black sea, scoring hits on military objectives and causing several big fires.

A 10,000-ton German transport was declared sunk in the Baltic.

Affray Stabbing Fatal to Man

August Bishop, 38, 223 South Laredo street, died in Robert B. Green hospital early Saturday from stab wounds suffered in an affray in the 200 block of South Santa Rosa avenue Wednesday.

The wound which caused his death was in the left side, police said. A suspect is being held.

TO HOLD MARKET SHOW.

CHICAGO, Oct. 24.—Officials preparing the Chicago markets show, to be held December 2-5, have ruled that only entries destined for market will be shown to eliminate the necessity of a double haul on the nation's war-taxed transportation facilities. The market show will be held this year to supplant the customary international livestock exposition which was cancelled because of transportation difficulties.

Missing '18 Ace Hunted In Pacific

HONOLULU, Oct. 24.—(AP)—American war birds of '42 were in the sky over distant Pacific waters today searching the sea for Capt. Eddie Rickenbacker, ace eagle of '18.

The war department said Rickenbacker, on an inspection trip for Lieut. Gen. H. H. Arnold, army air force commander, had not been heard from since Wednesday evening when he reported only one hour's supply of gasoline remained in his plane's tanks.

At that time, Rickenbacker and the crew of a large military plane were flying between Oahu, Hawaii, and another island.

United States army Hawaiian department headquarters announced every available plane and surface craft from the southwest Hawaiian islands was searching for the missing fliers.

PLANE CARRIES 10

The army did not disclose names of others aboard the plane, but it was believed here there were at least 10 persons, including the crew.

Normal equipment of all such planes includes rubber life rafts. These fragile craft have saved the lives of many military fliers forced down at sea in Pacific fighting.

The 52-year-old veteran, as America's first great ace, bagged 21 German planes and four enemy balloons. Dewey Rickenbacker, his brother, said at Beverly Hills, Calif., "this just isn't Eddie's time. After all, this isn't the first time he's been in a tight spot."

But Mrs. Elizabeth Rickenbacker, 79, the mother, had her misgivings when the captain visited her at Beverly Hills recently.

LAUGHS OFF FEARS

"My mother felt that something was going to happen to Eddie and asked him not to go," Dewey Rickenbacker said. "But he just laughed."

This optimism and fortitude carried Rickenbacker through a tough experience nearly two years ago when he was injured seriously in the crash of an airliner near Atlanta. Rickenbacker spent weeks in a hospital. Others among the passengers and crew of that plane were killed.

In New York, Rickenbacker's wife waited by the telephone.

Rickenbacker's trip over the Pacific was for a purpose similar to his recent inspection flight to England.

Nazis to Sing New Party Songs

LONDON, Oct. 24.—(AP)—The German people have been ordered to learn by heart and give lusty voice to a series of Nazi-sponsored songs, the British radio reported today, including one entitled "Siehst Du Im Osten Das Morgenrot?"—Do You See the Rosy Morn in the East

Another ditty listed was "Nach Ostland Wollen Wir Fahren"—Towards the East We Want to Drive.

The British broadcast, heard here by C. B. S., said Bernhard Rust, German minister of education, was entrusted with the task of getting the people in a singing mood.

R.A.F. Visits Italy 2nd Night in Row

LONDON, Oct. 24.—(AP)—The R. A. F.'s biggest bombers, back in force over Italy for the second successive night, showered explosives on Turin and Genoa again last night and also on Savona in a sweep over three of Italy's northernmost provinces.

Especially at Savona, a port on the Ligurian coast about 25 miles west of Genoa and one of the chief Italian foundry centers, the Fascist communique mentioned "notable" damage.

Italian foundry centers, the Fascist the British again said the raiding armada was a "strong force" and that all but three returned safely.

As on the preceding night, the British said, these included the R. A. F.'s great Stirlings, Halifaxes and Lancasters as well as the smaller Manchesters.

FLY 1500 MILES

All had to fly roughly 1600 miles round trip, virtually the entire distance over hostile territory, and twice cross the Alps. One Lancaster, they said, "flew most of the way out and all the way home on three engines."

The three bombed cities all are important as industrial and military centers and the Italians acknowledged all were pounded with explosives and incendiary bombs.

The Italian communique said the damage in Genoa and Turin, both heavily raked the night before, was not serious, but that in Savona "the damage was more notable."

CASUALTIES TOLD

The British caused no casualties in last night's raid on Genoa, the Italian communique said, but a large number of casualties was caused by "excessive crowding in a shelter". On the previous night, 29 were killed and 121 injured at Genoa, the communique said.

The communique reported one person killed and 10 injured at Turin last night and 14 killed and 46 injured between Savona and Vado Ligure, a small seaport three miles southwest of Savona.

Nazi-occupied France, the Netherlands and Germany were attacked yesterday by daylight. Three planes were lost.

French Africa's Governors Called

NEW YORK, Oct. 24.—(AP)—The British radio relayed a Nazi broadcast from Paris today saying a council of all the governors of French Africa had been called to meet at Dakar. The B. B. C., heard here by C. B. S., said the date was not given.

Chinese Repulse Jap Motor Column

CHUNGKING, Oct. 24.—(AP)—A Jap motorized column was thrown back after attempting to advance from Paotow in the northern province of Suiyuan a week ago, the Chinese high command announced today.

WADE ASHORE FOR BATTLE

Marine reinforcements for the impending all-out battle for the Solomons wade ashore on Guadalcanal island to reinforce comrades who seized the airfield there in August. —Int'l.

1,300,000 in Navy Forces

WASHINGTON, Oct. 24.—(AP)—The United States naval services—some 1,300,000 strong — today are steadily strengthening the nation's coastal defenses at home even as they increase their striking power on the high seas.

And a "very steady and very satisfactory rate of enlistment" is providing the men to do the dual job, says Navy Secretary Knox.

The secretary disclosed at a press conference yesterday the strength of the navy proper was nearly 1,000,000, and that there were approximately 200,000 in the Marine corps and about 110,000 in the regular coast guard.

Gandhi Wife's Death Reports Denied

BOMBAY, Oct. 24.—(AP)—Official quarters declared today there was no foundation for German radio reports of the death of Mrs. Gandhi, wife of India's imprisoned Nationalist leader.

(The German broadcast, heard yesterday in London, gave no details.)

Ship Named for Marine Launched

NEW YORK, Oct. 24.—(AP)—The destroyer Daly was launched at Staten Island today, named for the marine sergeant major who the navy said led a World war I attack in France with the cry, "Come on, you ——, do you want to live forever?"

Japs Closing in On Guadalcanal

WASHINGTON, Oct. 24.—(AP)—Jap forces, dealt a severe naval loss by Allied bombers, apparently were closing in on embattled Guadalcanal island today for an all-out assault on the American airbase.

Disclosing the Japs had landed on Russell island, about 30 miles northwest of Guadalcanal, naval spokesman acknowledged the invaders might have established similar positions on other neighboring small islands.

While the navy reported yesterday additional destruction of 12 Nip planes and repulse of a second enemy "feeler" attack on the air field, heavy bombers under General MacArthur sank or badly damaged 10 Jap vessels including a cruiser.

Raiding with "deadly effect," Allied headquarters said, the bombers dumped explosives on the cruiser, a destroyer and eight transports or cargo ships at Rabaul, New Britain, just west of Bougainville island in the Solomons.

The low-level attack before dawn yesterday (Australian time) was believed to have sunk the cruiser, destroyer and two large merchant vessels while severely damaging the other ships.

The Jap flank attack on troops holding Guadalcanal air field occurred October 21 and was the second in two days to be driven back. The maneuver was regarded as a test of the American force's strength, but if successful might have been followed by heavy waves of enemy soldiers.

On the same day, seven enemy bombers with a fighter escort sought to bomb the air field.

One bomber was shot down by anti-aircraft batteries and two others damaged. Grumman Wildcat fighters intercepted the Jap fighter planes and destroyed six, the navy added, at the cost of two Wildcats.

Several groups of enemy planes renewed the attack the next day. Flying conditions were poor, the navy said, but the Wildcats shot down an entire attacking group of five bombers.

American positions on Espiritu Santo island in the New Hebrides were shelled the night of October 22 by an enemy ship, presumably a submarine.

The Allied bombing raid on Jap naval concentrations was the second in two days. The planes from Australian bases previously dropped 10 tons of bombs on Jap vessels in Bougainville's Buin harbor.

Aerial attacks since the strong enemy naval forces were sighted about 10 days ago thus far have sunk or damaged five Jap cruisers, 10 destroyers and 14 other ships, mostly transports.

Meantime, the unsuccessful Jap drive for the Allied Port Moresby base in New Guinea was subjected to a continued Australian offensive which has succeeded in pushing enemy troops back toward their Kokoda invasion base.

Greeks Hold Sector On Egyptian Front

NEW YORK, Oct. 24.—(AP)—Greek army units "have been entrusted to hold an extremely important sector of El Alamein front in Egypt," the British radio said today, quoting Panayotis Kannelopoulos, vice president of the Greek government in exile.

The B. B. C., heard here by C. B. S., said the Greek official had just returned from a visit to Greek forces and that he found them "in fine fettle and ready to fight anywhere."

Talk Is Cheap, by This Lad's Method

MINEOLA, N. Y., Oct. 24.—(AP)—A war plant riveter has a girl in Miami, Fla., he's been phoning five nights a week for four weeks.

It cost him only $5, he said in court yesterday. That was the price of a big bag of slugs. His 30-day sentence was suspended.

A phone company representative urged the suspended sentence, saying the worker must have learned his lesson and riveters are too important these days to be in jail.

Forecast

San Antonio and vicinity: Little change in temperature Saturday afternoon and night.

East Texas (east one-hundredth meridian): Little change in temperature Saturday afternoon and night. Light local showers near the coast Saturday afternoon.

(Weather bureau data on Page 2)

Rommel Hit By Sky, Sea And Land

CAIRO, Oct. 24.—(AP)—Britain's rebuilt and refreshed Eighth army charged into the Axis' El Alamein line today in an offensive sprung in the night with blows against the enemy by land, sea and air.

The Allies thus beat the Axis African corps to the punch and launched what may be the battle to decide the fate of the Mediterranean this winter.

CLOSE CO-ORDINATION

With all branches in close coordination, the armored army of the desert thrust forward under strong air support, including planes and fliers from the United States, while a task force of the Mediterranean fleet struck deep on the enemy's sea-girt northern flank near the Egyptian port of Matruh.

A naval communique issued at Alexandria said the naval force suffered no casualties and only "superficial damage" to one boat despite an enemy air attack.

U. S. air forces fighter planes were active in the final preliminaries of the long aerial preparation for the offensive, raking the Axis' forward landing ground yesterday and escorting Allied fighter-bombers on similar missions.

The attack started under a full desert moon last night, but it still was too early to discern any trend.

HURL SHELLS BY THOUSANDS

Under a strong cover of warplanes, the Allies lunged against the enemy positions in a swirl of sandy dust with tanks and guns—many of them American-made—hurling thousands of shells.

The main battle was concentrated in a comparatively small area, streching 40 miles inland from the coast to the Qattara depression, but probably it will spread rapidly as the British have been sending desert patrols hundreds of miles across the desert to strike at the rear of Axis Field Marshal Rommel.

Lieut. Gen. B. L. Montgomery, Commander of the British Eighth army—a collection of troops from various parts of the empire, as well as Poles, Free French and other Allies—announced recently that his men were "fighting fit and ready for anything."

The general had been faced with the task of replacing the "upwards of 50,000 men" which Prime Minister Churchill announced had been lost along with valuable equipment in the retreat from Libya to the El Alamein line in June. The El Alamein line was stabilized on July 1.

ALEXANDER COMMANDS

The British offensive was under the over-all direction of Gen. Sir Harold Alexander.

Alexander, 50-year-old hero of Dunkerque, succeeded Gen. Sir Claude Auchinleck as Middle East commander August 18.

Days of steady air attacks by British and United States fliers upon German-Italian air fields and communications opened the way for the ground push.

"Fierce fighting developed and is continuing," said a terse communique issued jointly by British headquarters and the R. A. F.

Thus the Allies got the jump on Rommel, who boasted in Berlin recently that he would crush their forces in the Middle East.

2 U. S. Ships Lost In Russia Convoy

WASHINGTON, Oct. 24.—(AP)—Two medium sized United States merchant vessels have been sunk in the North Atlantic as a result of an enemy aerial torpedo attack, the navy said today.

The vessels were in a convoy en route to a Russian port

Survivors of the attack, which occurred about the middle of September, have landed at an east coast port of the United States.

City Health Chief Goes to Convention

Dr. W. A. King, city health officer, Saturday was en route to St. Louis, where he will attend a convention of the American Public Health association Monday through Friday. King is a member of the board.

While London Crowds Cheer

British Royalty Fetes Eleanor

LONDON, Oct. 24.—(AP)—Mrs. Franklin D. Roosevelt arose early today and breakfasted alone in her apartment at Buckingham palace as a crowd of sightseers, including many American soldiers, gathered outside the palace.

Mrs. Roosevelt's apartment was one of those damaged by bomb blasts some time ago and its windows are partly boarded over.

After leaving the king and queen, she spent a busy morning conferring on plans for her tour of Britain before leaving to give a press interview. The president's wife hopes to study the part which British women are playing in the war and visit American troops.

PRINCESSES ABSENT

The two princesses were not in the palace when Mrs. Roosevelt arose this morning. They stayed in London yesterday until she had arrived by train after her trans-Atlantic flight and then went to their country home for the night. Mrs. Roosevelt had a long talk with them before they left.

Her press conference attended by more than 100 American and British reporters, was held at the United States ambassy.

In response to British questions, she said it was "hard to say" whether or when there would be conscription of women in the United States and she doubted whether prohibition would be brought back.

LUNCHES AT PALACE

With King George present, Queen Elizabeth entertained at a Buckingham palace luncheon today. Mrs. Roosevelt talked informally to the guests who included the heads of major women's military and civilian war services such as the ATS, WRENS, and WAAFS. Mrs. Oveta Culp Hobby, director of the American WAACS, attended the luncheon.

She was greeted personally by King George VI and Queen Elizabeth yesterday on her arrival in London.

To make the welcome complete, Mrs. Roosevelt's son, Elliott, a lieutenant-colonel with the U. S. army air forces in Britain, dined with the royal family last night.

CHURCHILL ATTENDS

Among the dinner guests were Prime Minister and Mrs. Churchill, U. S. Ambassador John G. Winant, Field Marshal Jan Christiaan Smuts, premier of the Union of South Africa; his son, Capt. Jacob Smuts; Lord Louis Mountbatten, head of the Commandos; Lady Mountbatten, and Mrs. Roosevelt's Secretary, Malvina Thompson.

Mrs. Roosevelt sat next to the king at dinner, which was described as a simple meal.

After the meal the guests were entertained by a showing of Noel Coward's film of the British navy, "In Which We Serve."

She was accompanied to Britain by Mrs. Oveta Culp Hobby, director of the WAACS.

PRESS PLEASED

London morning newspapers did their part to extend a hearty welcome to the president's wife.

Typical was a headline in the Daily Mirror, "We're Sure Glad to Meet You, Ma'am", and an editorial in the Daily Express declaring: "You are most welcome, madam, and at any time in the past you would have been most welcome, too."

The Times, after expressing pleasure at Mrs. Roosevelt's arrival, said:

"We shall expect the results of a searching but thorough observation of England at war will ultimately find their way to the president, whose insight into our affairs cannot be too penetrating for our desire."

PART 1

The Deal of the Century
San Antonio, 1942

The front page of the October 24, 1942, edition of the *San Antonio Light* showed American marines wading to shore for an "all-out battle for the Solomons," and two headlines were emblazoned across the top of the page. One announced that the United States and Britain had opened a battlefront in Egypt; the other reported that the San Antonio Public Service Company had been purchased by the City of San Antonio.

Other news of World War II filled the page, describing two nights of bombing in Italy by Britain's Royal Air Force and attacks on Nazi Germany's field marshal Rommel "By Sky, Sea, and Land." Not quite a year after the Japanese attack on Pearl Harbor, battles were raging in Europe, Africa, and the Pacific. The world watched with dread and wonder as planes and tanks and men went into combat, with newspapers and newsreels documenting the drama.

Despite news of the war, other big news in San Antonio was the talk of the town that day. Mayor Charles Kennon Quin held a press conference on

Saturday morning, in time to make the afternoon edition of the newspaper, announcing that the City had made a deal for providing power to San Antonio. "Full details of the deal will be made public just as soon as we can possibly do so," the mayor said, indicating that it "might result in some saving to San Antonio." This prediction would prove to be the understatement of the century.

Mayor Charles Kennon Quin worked with city leaders to purchase the San Antonio Public Service Company from American Light & Traction in 1942.

Prior to the purchase, San Antonio received its gas, electric, and transportation service from San Antonio Public Service Company (SAPSCo), which also provided electric service to the rest of Bexar County and parts of nine neighboring counties, including the towns of Boerne, Hondo, and Floresville. There were three existing power plants: Station A, built in 1909 on Villita Street in downtown San Antonio; Station B, built in 1917 on Mission Road; and the Comal plant, built in 1926 just outside Landa Park in New Braunfels.

For the past decade, construction in San Antonio had been booming. Military presence had increased at Fort Sam Houston Army Base, and Air Corps personnel training had intensified at Randolph and Kelly air fields in anticipation of World War II. San Antonio's population had surpassed 250,000, and as the city grew SAPSCo began acquiring smaller electric generating companies, including the Travis plant, which served the St. Anthony Hotel, the world's first hotel to have a central system for air-conditioning and heat.

Three power plants provided energy to San Antonio in 1942. The newest was the Comal plant in New Braunfels, built in 1926.

Not unexpectedly, with the growing demand for power, large national companies had acquired utility companies in many parts of the country. The United Light & Power Company had controlled SAPSCo since 1930, as well as companies operating in Michigan, Iowa, and several other states. This large holding company was owned by the New York–based American Light & Traction Company, part of the Wall Street banking house Emerson McMillin.

In the late 1800s Emerson McMillin recognized that gas and electricity were the way of the future, and by 1900 he had focused the banking house that carried his name on mergers and acquisitions in that new realm of business, using a holding company he called American Light & Traction, where he served as chairman of the board. Within a year the company controlled over forty small local power suppliers that produced gas and electricity, along with streetcar properties, in various parts of the United States.

McMillin's rise to wealth and influence is one of the great American success stories. Born in 1844, he grew up in poverty in Ewington, Ohio, the youngest son of an iron furnace manager. At the age of twelve he became an apprentice in this rough and tumble industry alongside his father and five brothers. Strong and unafraid of hard physical work, he quickly mastered the mechanics of the blast furnace engines and the workings of a charcoal kiln. He did not go to school until he was fifteen, and two years later he joined his father and older brothers in the Union Army to fight in the Civil War. Known as "the Fighting McMillins," two brothers were killed and the youngest McMillin was wounded five times.

Banking house and holding company mogul Emerson McMillin got his start working in the Ohio iron foundries in the mid-1800s.

While his regiment was camped in the mountains of Virginia, McMillin became interested in geology and the natural sciences; his explorations and the knowledge he obtained there would eventually earn him recognition as one of the foremost authorities in the gas world and lead him to the highest financial circles on Wall Street.

After the war McMillin and a brother purchased and sold coal, comfortable back in the world of blast furnaces, saw mills, and similar industries. He drifted into the town of Ironton, Ohio, where a new gasworks was being built. Hired as a laborer by the superintendent, McMillin watched every phase of the construction and manufacturing process as he worked, and when the project was complete he was encouraged to stay. The superintendent told him that the gas industry was in an experimental stage, that it was a field filled with opportunities that might never come again, and McMillin heeded that prediction. By 1867 he was superintendent of the Ironton Gas Light Company, and he focused on learning everything about metallurgy and the generation of gas, installing a research lab at the plant.

His in-depth knowledge catapulted him to the top of the new field, and in rapid succession over the next decade he became general manager of Lawrence Iron Works, vice president and general manager of the Crescent Iron Company, president of the Iron & Steel Company, and general manager of the New York & Ohio Steel Company, which operated rolling mills, blast furnaces, and

McMillin's offices on Wall Street were described as "the most luxuriant in the country" in 1897.

coal and iron mines. By 1883 he had purchased several small coal gas plants in various locations and had completed a takeover of the Columbus Gas Works. In 1889 he combined four warring gas companies in St. Louis to form Laclede Gas Light Company, which eventually became the largest natural gas distribution utility in Missouri, operating today as Laclede Gas.

McMillin was also an agent for the London-based American Industrial Syndicate, Limited, along with New Yorker George Shepherd Page. Sir Julian Goldsmith, president of more than a hundred gas companies and one of the richest men in England, was a director of the organization, along with six other wealthy British capitalists. The syndicate was seeking investments in American gas properties; McMillin created a massive $13.5 million deal with the syndicate and Laclede Gas Light that earned him a door-plate on Wall Street bearing the name Emerson McMillin & Company, Bankers. The deals got even bigger and included the successful launch of the East River Gas Company of Long Island City, which supplied gas to New York City through a tunnel under the East River. An article written in 1897 by Lila Rose M'Cabe for the *Ironton Register* described the eleven-room banking house of Emerson McMillin as "the most luxuriant in the country, with oak walls, polished wood floors, marble tile corridors, Turkish rugs—they replete the paraphernalia that goes to make up the business shrine of a modern millionaire." It was in these opulent offices that the City of San Antonio would negotiate, two decades after McMillin's death, to create the best deal in its history.

In 1935 Congress passed the Public Utility Holding Company Act to discourage control of the electric utility industry by a few large corporations. President Franklin Roosevelt pushed hard

against public utilities following his election in 1932, and the legislation provided a way to take action against them. The law's impact was not felt in San Antonio until 1941, when American Light & Traction was ordered to divest itself of SAPSCo and certain other South Texas holdings.

Entrepreneurial business leaders in San Antonio saw an opportunity in the divesture and urged Mayor Quin and the city commissioners to negotiate the purchase of SAPSCo. Wilbur L. Matthews was the principal attorney for the utilities company and a senior partner at Brooks, Napier, Brown & Matthews, one of the city's earliest law firms, first established as Ogden, Brooks & Napier in 1904. Matthews had begun representing SAPSCo in 1929 when senior partner Walter P. Napier became president of Alamo National Bank. In his 1983 book *San Antonio Lawyer*, Matthews recalls that many others recognized

State Sen. Alvin J. Wirtz led efforts for the Guadalupe-Blanco River Authority to purchase SAPSCo.

the divesture as a potential "gold mine" as well, including the Guadalupe-Blanco River Authority (GBRA), based in Seguin.

Texas State Senator Alvin J. Wirtz was from Seguin, and he was a close friend of U.S. Congressman Lyndon B. Johnson, who was "in favor" with President Roosevelt. With these connections, Wirtz had been successful in securing government aid for financing dams on the Colorado River and in establishing the Guadalupe-Blanco River Authority.

American Light & Traction's upcoming divesture presented a way to parlay his political relationships into an opportunity for the GBRA. In February 1942 Wirtz traveled to New York to meet with William H. Woolfolk, president of American Light & Traction. He proposed that the GBRA would buy the capital stock of SAPSCo, dissolve that company, retain the electric system, and dispose of the gas and transportation system by selling it to the City of San Antonio or another company. The money to buy the stock would come through the issuance of $42 million in bonds by the GBRA. Wirtz's plan envisioned that the San Antonio electric company outside of Bexar County would be operated by the Lower Colorado River Authority under a lease from the GBRA, and that the Lower Colorado River Authority would also acquire the electric properties of Central Power & Light, serving all of South Texas. When Matthews wrote his book, documenting some of his most interesting cases and clients, he described Wirtz's bold plan as "a Texas version of the Tennessee Valley Authority."

According to Matthews, Woolfolk told Wirtz that the stock of SAPSCo was worth more than $35 million, but Wirtz showed no interest in the proposal and returned to Texas without a deal. A

President Franklin D. Roosevelt (left) and U.S. Congressman Lyndon B. Johnson (right) supported Wirtz's push for the GBRA to purchase the utility company.

few months later Wirtz held a secret meeting with Quin and members of the City Commission at the Plaza Hotel in San Antonio. Much like the plan he had presented in New York, he proposed that the GBRA would acquire SAPSCo's stock and sell the electric, gas, and street transportation to the City of San Antonio for $35 million. Under his plan the GBRA would keep the Comal plant and the electric properties outside Bexar County without cost and would sell the electricity generated at the Comal plant back to San Antonio. The mayor and commissioners quickly rejected his proposal, and Wirtz left in anger, stating that "from now on we meet at arm's length."

SAPSCo officers and directors sided with the City and were opposed to Wirtz's proposal, of course, but they were not in a position to take action. Although the Chamber of Commerce, local newspapers, and citizen groups had been against municipal ownership when the divesture was announced, public opinion began to shift when it became known that Wirtz and the river authorities were seeking to take over San Antonio's utilities. The possibility that the future of San Antonio might be controlled from outside the city was not attractive.

Newspapers denounced the GBRA and nicknamed the organization the "Ghost of the Guadalupe," pointing to the fact that it was an organization without any assets, organized to build dams and other facilities to conserve water on two rivers that did not pass through San Antonio. Within a few weeks the prevailing attitude was that city ownership of the utilities was the lesser of two evils. An editorial in the June 6, 1942, *San Antonio Express-News* reported that public ownership appeared "inevitable" and called for three essential conditions in the final deal: a fair purchase price, payment by the issuance of revenue bonds, and management of the properties by a nonpolitical, self-perpetuating board of highly qualified members without political patronage.

Competition to buy SAPSCo intensified, and the commissioners adopted an ordinance calling for the issuance of $35 million in electric and gas system revenue bonds to be delivered on October

When the mayor and his city attorney, Van Henry Archer, traveled to Austin in 1942 to meet with the state's attorney general, their expenses totaled $7.11.

24, 1942. In the meantime the GBRA was seeking to issue about $42 million in revenue bonds to purchase the properties. The race was on.

On June 20 Mayor Quin and his city attorney and secretary, Van Henry Archer, traveled to Austin for a meeting with the state's attorney general, Gerald Mann, to discuss the situation. The minutes from the subsequent city commissioners meeting reflect that Archer was reimbursed $7.11 for his travel expenses from the City's general fund. The meeting was a success; Mann refused to approve the GBRA's bonds based on the fact that the organization was a water conservation organization not authorized to buy, own, or operate the utility properties. The GBRA filed an appeal with the Texas supreme court, and legal entanglements would linger for several years after the successful purchase was made by the City of San Antonio.

In the final months of negotiations Quin and Archer traveled much farther than Austin to cement the deal. Recognizing that time was critical in the final discussions with Woolfolk and American Light & Traction in New York, they chose to make the trip by plane instead of train.

When the deal was finalized, the purchase price was $33,990,000, financed entirely through a bond issued at a 2.854 percent interest rate. Because the City had an excellent credit rating, no cash was required. A trust indenture was set up as a contract between the City and the bondholders, establishing rules for operation of the new City Public Service Board (CPSB). The utility's revenues would be used first for operating, maintaining, and repairing the gas and electric system and for insuring the utility for materials, fuels for energy production, and salaries. The contract also provided for an annual tax equivalent payment of $210,800 to the City and an annual property tax payment of $113,700 to the San Antonio Independent School District.

The governing body was designed as a five-person board of trustees, one of whom would be the mayor of San Antonio. Trustee terms were five years, with eligibility for one reelection; and trustees would be paid $2,000 annually, except for the chairman, who would be paid $2,500. The board would fill vacancies by majority vote, elect its own chairman and vice chairman, and appoint the CPSB general manager. This arrangement, including stipend amounts, has worked well for seventy-five years and remains in place today.

Members named to the first governing body were Walter Napier, then chairman of the board of Alamo National Bank; Franz Groos, president of Groos National Bank; D. F. Youngblood, president of Southern Steel Company; Col. W. B.

Walter Napier, who had a long history with SAPSCo, was one of the first trustees appointed to the newly established City Public Service Board.

Tuttle, president of the utility company about to be purchased; and Mayor Quin.

Napier became a partner at one of San Antonio's earliest law firms, which became Ogden, Brooks & Napier in 1908. He had represented SAPSCo before he served as the first president of Alamo National Bank, and his protégé Wilbur Matthews became the utility company's attorney. Napier eventually became chairman of the bank, as well as serving as chairman of the San Antonio Chamber of Commerce in 1939. Youngblood was the founder of Southern Steel, established in 1897, on South Presa Street. In addition to manufacturing steel products, mainly for prisons, he held several patents for liquefied gas storage and dispensing systems. Groos came from an illustrious banking family, originally from Eagle Pass, Texas. Frederick Groos, his uncle,

Col. William B. Tuttle (front row, sixth from left), pictured with CPSB employees, was named both a trustee and general manager of the utility company.

established F. Groos & Company in 1854, probably the state's first banking business. In the early days it focused on private mercantile lending, but as other banking services were introduced it relocated to San Antonio in 1874 and became Groos National Bank. In 1894 young Franz graduated high school from the San Antonio Academy, at a time when every graduate received acceptance to the University of Pennsylvania and Harvard, Yale, and Princeton Universities without having to take entrance examinations. Groos chose Princeton and returned to San Antonio after graduation to go into the family business. He was known for combining astute lending practices with kindness.

As SAPSCo's principal attorney, Matthews attended the closing, which he described as one of the most amazing he had ever seen. "I know of no 'closing' in which as many different transactions occurred and as many parties were involved with deliveries of hundreds of multipage documents, all in a single four-hour period, in the offices of Dewar, Robertson & Pancoast on the third floor of the old National Bank of Commerce Building on Main Plaza," he wrote. It's safe to assume that when Mayor Quin predicted that the nearly $34 million deal might save San Antonio some money, no one imagined that over the next seventy-five years it would provide

In 1942 San Antonio had a population of 250,000 and was rapidly growing.

more than $7 billion in revenues to the City—as well as power.

The municipally owned CPSB held its first meeting the next day, ushering in a new era for utilities that would have far-reaching impacts on San Antonio. Its first order of business on October 25, 1942, was to elect a chairman.

Col. William Bockhout Tuttle, former president of SAPSCo, was elected to the position and appointed general manager of the company the same day. Like Emerson McMillin, Tuttle was

a native of Ohio. He had enjoyed a comfortable childhood and graduated from the University of Virginia with a mechanical engineering degree. He apprenticed at the Consolidated Gas Company of New Jersey, a subsidiary of American Light & Traction, and was promoted to general manager before he was sent to San Antonio in 1906 to evaluate the operating systems of the San Antonio Gas Company and the San Antonio Traction Company, both subsidiaries of American Light & Traction. Early the next year he returned to run the transportation company and to supervise construction of a plant on Salado Street for San Antonio Gas. Tuttle recognized that new landmarks like the San Antonio Public Library and the Southern Pacific Railroad's beautiful Sunset Station were solid indicators that the city was becoming more modern and sophisticated, and he embraced the community wholeheartedly.

By 1910 he was vice president and general manager of the traction company, which had replaced its mule-drawn, open-air cars of the late 1800s with fifty-four enclosed cars that ran along electrified lines—connecting the downtown area to San Pedro Park, the railroad stations, the growing neighborhoods of Monte Vista and Laurel Heights, and the fairgrounds south of town. Tuttle was active in the community from his earliest days there, volunteering on several civic boards, participating in the first River Parade in 1907, and serving as chairman of the Chamber of Commerce in 1913.

When transportation, gas, and electric companies finally merged into the San Antonio Public Service Company in 1917, Tuttle was there. During his first decade in his new hometown he had seen San Antonio's greatest proportional population boom, from about 45,000 in 1910 to more than 96,000 in 1920. Railroads and the streetcar industry had given mobility and access to its citizens, gas and electricity had brought light and heat, and buses and automobiles were beginning to change transportation still more. From 1920 to 1940 San Antonio boomed in every way, and by 1942 its population surpassed 250,000.

With thirty-seven years of experience in every aspect of the utilities business, Tuttle had seen firsthand the extraordinary opportunities that light, power, and transportation produced, helping San Antonio become one of the largest cities

Tuttle's experience with San Antonio's utility began in 1907, and by 1942 he was known as "Mr. Public Service."

in Texas. Nicknamed "Mr. Public Service," he was recognized for the major role he played in the city's modernization and growth, and when he accepted his new dual roles at the CPSB he shared Mayor Quin's confidence that the deal would create even more positive and important changes in San Antonio.

SAN ANTONIO DE BEXAR.

PART 2

Let There Be Light
San Antonio, 1858–1900

In the 1850s San Antonio was a remote place in the new state of Texas, dusty or muddy depending on the rains, still recovering from a terrible cholera epidemic that occurred in 1849. Water was provided by the acequias, stone waterways built during Spanish rule, and homes were lit by candlelight and oil lanterns—or sometimes not at all.

At the request of Galveston's Catholic bishop, John Mary Odin, four Marianist missionaries, part of the Brothers of the Society of Mary, were sent to Texas from France in 1852. After a short stay in Galveston, the state's burgeoning port city, they were given their assignment to establish a school in a rather wild place called San Antonio. They stepped off the Indianola stagecoach and saw a frontier town with dirt streets, cowboys, and a few buildings. They opened the school above a livery stable that summer with twelve students—all boys—and a faculty of five. Within a year the school's first director, Brother Andrew Edel, had increased enrollment to one hundred boys and had relocated the school to

its own two-story building. Aware of the success of St. Mary's Institute, two free public schools opened in San Antonio in 1853, one for boys and one for girls. Over the years St. Mary's Institute would become St. Mary's College and later San Antonio's oldest university.

In the mid-1800s San Antonio was a dusty town that went dark after sunset.

The Menger Hotel opened with great fanfare in 1859. A few years later it added gaslights, and electricity would follow two decades after that.

A few years later German businessman William A. Menger opened a hand-operated brewery south of the Alamo. Established in 1855, Menger's Western Brewery is usually considered the state's first commercial brewery. Its cellars were chilled naturally by the Alamo Madre ditch that once flowed through the brewery, and four workers operated the brewing equipment to produce lager beer by hand. When Menger realized that his guests often needed a place to sleep, he and his wife, Mary, expanded into an elaborate two-story, fifty-room hotel that opened with great fanfare in 1859. Although it ushered in an era of sophistication on the Texas frontier, the hotel would not provide gas light for several more years, followed by water and electricity two decades later. The same year that the Menger Hotel brought its luxury to San Antonio, the Fifth Avenue Hotel opened in New York City. Faced with white marble, it stood five stories tall and boasted the country's first hotel passenger elevator, powered by a steam engine. A correspondent for the *Times* of London, in New York to cover the visit of the Prince of Wales in 1860, called the hotel "a larger and more handsome building than Buckingham Palace." While the Menger could not match the Fifth Avenue's glamour, it pioneered the hotel industry in San Antonio.

More sophisticated cities in the eastern United States, like Baltimore, Philadelphia, New York, and Boston, already had streets well-lit by gas lamps by the early 1800s, and wealthier citizens had replaced their oil lamps at home with this modern technology as well. The first gas streetlights went into service in Baltimore in 1816, and lamps came to life in New York in 1825, from Canal Street to the Battery, lit by gas provided by the New York Gas Light Company, the earliest predecessor of Consolidated Edison, which would

The San Antonio Gas Company (chimney on left), established in 1860, was located on the banks of San Pedro Creek.

become one of the country's largest investor-owned energy companies.

It wasn't until later in the decade that an entrepreneurial San Antonio businessman, S. R. Dickson, amassed what was a fortune in those days—$120,000—and applied to the City of San Antonio for a franchise to manufacture and distribute gas for streetlights. He received an ordinance authorizing him to proceed, stipulating that "three miles of pipes shall be laid in all, through the streets, lanes, alleys, and public grounds, and shall be laid within three years in order to provide streetlights for the City." His San Antonio Gas Company began service on January 26, 1860, with a plant on the corner of Houston Street and San

Pedro Creek. Gas was distilled by burning wood resin, the best illuminant of the times, shipped from New England to the Gulf Coast port of Indianola and hauled to San Antonio by oxcart. Within a year Main Plaza and numerous saloons in the town of 8,000 had the modern glow of gaslights.

The city had its first bank soon after, when Col. T. C. Frost founded Frost Bank in 1868. A few years later, in 1873, the Sundry Civil Service Bill included $100,000 for an expanded army post, to be built on ninety-three acres of land deeded by the city. A small garrison and quartermaster supply depot had been operating in San Antonio since 1845, located in rented buildings in

GOVERMENT TOWER

Construction of a new army post began in 1876 while the Indian Wars still raged.
The clock tower, eighty-seven feet high, dramatically changed the city skyline.

the Alamo compound, and plans for a much bigger post brought a sense of security to a town that still felt the threat of the violent Indian Wars. The Edward Braden Construction Company won the contract to build the post for $83,900, and work began in 1876 under the supervision of Maj. Gen. Edward Ord, a West Point–trained army engineer. The modern quadrangle, designed by George Henry Griebel, provided an eighty-seven-foot watchtower and a 6,400-gallon water tank, and the first troops moved in the following year. When Apache chief Geronimo was captured a few years later, he was housed in tents just west of the quadrangle until he could be moved to a reservation.

Renowned architect Alfred Giles designed fifteen officers' quarters in 1880, a parade ground was added, and over the next few years the Post at San Antonio expanded to include infantry, cavalry, and light artillery. In 1890 it was designated Fort Sam Houston, named for the American soldier and politician who fought for Texas independence from Mexico, served as president of the Republic of Texas, was a major force in bringing Texas into the United States, was elected to the U.S. Senate, and served as governor of Texas.

When the Southern Pacific Railroad came to San Antonio in 1877, ranchers could move their cattle without the long drives of the past, farmers could get their cotton to market cheaply, and perhaps most impactful, anthracite coal from Pennsylvania could be transported easily, revolutionizing the production of manufactured gas.

The Apache chief Geronimo and some of his warriors were captured and housed at Fort Sam Houston until they were moved to a reservation.

Lower in heat value but much less expensive than resin, coal was used in restaurants and hotels and eventually in homes for cooking, heating water, and light. Delivery of power changed drastically. Railroads brought real modernization to town; the first passenger train arrived in 1878, and by 1880 the city's population had grown to nearly 20,000.

Just a year after the first locomotive pulled into San Antonio, the town started a public transportation system consisting of a few mule-drawn streetcars that traveled on narrow tracks from downtown to San Pedro Springs, a favorite recreation area about three miles away. Col. Augustus Belknap, an enterprising New Yorker, had acquired the rights to construct tracks

San Pedro Park was a favorite recreation spot in the late 1800s.

connecting Main Plaza to San Pedro Springs. He had seen the success of streetcars in New York starting in 1832, with a route that stretched from downtown to Harlem. Forty-five years later he thought San Antonio was ripe for the service. Though the park and springs were a significant draw, this alone could not justify the cost of a streetcar line. Belknap knew installation of the line would lead landowners to sell individual lots along San Pedro Avenue, which ultimately would provide a clientele base. The original streetcars consisted of a mule or a horse, a driver, and a car body that traveled on rail track. Each car carried fifteen passengers, traveled at five miles an hour, and cost six and a half cents to ride. As the system

expanded from Houston Street to Main Plaza, neighborhoods developed along the route. Over the next five years more track was laid and another streetcar system connected the three major plazas, two railroad stations, the San Antonio Arsenal, and San Pedro Springs.

In 1877 a new waterworks began to pump fresh water from the San Antonio River, but it would be another decade before water service was available to the majority of homes and San Antonio would have its first fire company. Mayor Francois P. Giraud had been accepting bids for a company to supply city water from its river since 1873. The contract was awarded to Jean Baptiste LaCoste to operate the San Antonio Water

A public transportation system, comprised of mule-drawn streetcars, offered a route from Alamo Plaza to San Pedro Springs in 1879.

Works Company. Located near the headwaters of the San Antonio River in what later became Brackenridge Park, a water pressure–operated pump lifted water to a reservoir in the present-day San Antonio Botanical Garden. The site was high enough for water to flow by gravity into the distribution system. Col. George Washington Brackenridge was a major financial backer of the company. By 1879 he had become the president, and by 1883 he held the controlling interest. The company changed hands several times over next few decades, eventually becoming the City Water Board in 1925.

Brackenridge came to Texas from Indiana in 1850, already an accomplished young man with training as a surveyor and engineer and a law degree from Harvard University. He worked as a salesman in Port Lavaca and was so successful that he established a mercantile business in 1853 in the small town of Texana where his family joined him. During the Civil War he became immensely wealthy through profiteering, buying cotton directly from growers and shipping it out of Matamoros to New York and on to England and France. After claiming Union sympathies, he was forced out of Texas for a few years and moved to Washington, D.C., where President Abraham Lincoln appointed him U.S. treasury agent in 1863. Ironically he was dispatched to Mexico the next year to try to persuade President Benito Juárez to cease that country's cotton trade with the Confederacy. When the Civil War ended Brackenridge returned to Texas and settled in San Antonio.

The Brackenridge family settled in San Antonio in the late 1800s and played a major role in the city's economic and cultural development.

Meanwhile, in a laboratory in Menlo Park, New Jersey, inventor Thomas Edison and his team of researchers had been working to develop the incandescent light bulb, which used a long-burning filament that promised to make electric light commercially viable at last. Edison succeeded in 1878, changing the way the world lived after sunset. While the discovery of electricity goes back as far as Benjamin Franklin and his famous kite experiment, harnessing the power in safe and affordable ways had been elusive.

The incandescent light bulb changed that, and in 1881 an enterprising group of San Antonio businessmen founded the San Antonio Electric Company. It began operating the next year, and its generating equipment was housed in the same building as the Nic Tengg printing company, at 31 West Commerce, across the street from the San Antonio National Bank Building. The first equipment consisted of a fifteen-horsepower steam engine belted to a dynamo capable of operating ten large arc lights. According to a newspaper article about the equipment, "the machine is arranged so that it is self-oiling and will run for days without attention except at the moment of starting and stopping. The carbons are equally pointed and hence the light radiates in all directions. This light is therefore equally well adapted to low and high ceiling rooms and also the most practical system for outdoor illumination."

The company's backers promised that the system "would place the most important part of the city, during the darkest nights, in almost as clear light as that of the mid-day sun and thus do much toward insuring against those evils, losses and abuses which the shroud of midnight tends to invite rather than dispel." In a frontier town with no paved streets, where bars outnumbered grocery stores and gambling and gunfights were commonplace, the promise of light was extraordinary and exciting.

On a chilly day in March 1882, workmen installed a strange-looking contraption on a pole in front of the U.S. Post Office on Alamo Plaza. The men finished their job at dusk and inserted a plug into the machine. Suddenly the post office was diffused in light, and the crowd that had gathered to watch cheered in amazement. In the week that followed, two of the city's most popular saloons had electric lights installed as well.

Thomas Edison's invention of the incandescent light bulb in 1878 made electric light commercially viable.

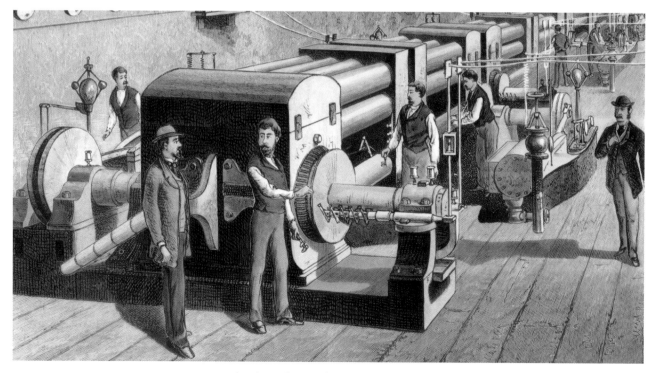

Edison's Pearl Street plant in New York operated with one generator, a belt-driven dynamo.

The U.S. Post Office on Alamo Plaza was lit by electricity in March 1882, six months before the Edison Illuminating Company of New York provided service to customers in lower Manhattan.

It would be another six months before New York provided the electricity for fifty-nine customers in lower Manhattan, generated by Edison Illuminating Company of New York's new power plant on Pearl Street and its one dynamo, a belt-driven machine that transmitted direct current over very short distances.

By the end of the month San Antonio's *Evening Light* newspaper reported that "as soon as the cost of electric lighting is lowered a little, it will be considered a prime necessity by every merchant who pretends to keep pace with the times." The Menger Hotel installed the new arc lights, providing a comparison to the gaslights that were still in vogue. The Vaudeville Variety Theater, owned by an Irishman named Jack Harris, was another of the businesses to keep pace. But despite the lights he installed in 1882, a tragedy struck the saloon that reminded citizens that San Antonio was still a "cowboy town," dangerous after dark.

In 1882 the Vaudeville Variety Theater was one of San Antonio's first commercial establishments to install electric lighting.

Harris had fought during the Civil War in the Confederate Army, where he served with Ben Thompson and King Fisher, both gunmen of the Old West. In 1868 he joined the San Antonio police force and a few years later established a saloon with Joe Foster. The Vaudeville quickly became the most popular watering hole in town. In those days there were more saloons than general stores or banks, and often they were owned by police officers. They were ornately furnished, served liquor, and sometimes offered gambling and dance hall girls—and no respectable woman dared enter them.

In 1880 Thompson spent some time at the Vaudeville, gambling heavily with Harris's partner, Joe Foster. Thompson lost badly, left the saloon angry, and returned to Austin. Ironically he too owned a saloon—the Iron Front—and served as that city's chief of police. Harris and Foster sent word to Thompson that he was no longer welcome at their establishment, but in July Thompson returned wearing his guns. According to witnesses, he avoided the saloon's lighting, stood in the dim shadows of the entry, killed Harris, and went to the Menger Hotel to spend the night. The drama increased when, after turning himself in to San Antonio police, Thompson was acquitted a few months later. Many of Harris's friends vowed revenge, and Thompson knew he was unpopular in San Antonio.

A few years later he dared to return. He met his friend and fellow gunman King Fisher at the Menger. By now the legendary hotel had added a wing, expanded its dining room, and installed electric lighting in some areas. Water was piped into every room, providing guests with the luxury of private bathrooms. A bar, billiard hall, and barber-shop were connected to the hotel, and the Menger was a favorite meeting place for visitors and locals. Thompson and Fisher had drinks in the bar, attended a play at the Turner Opera House, and later went to the Vaudeville. The facts are murky, but at some point there was a confrontation with Harris's old partner and several saloon employees, who ambushed the two gunmen, killing them in a hail of bullets. No action was ever taken against them. San Antonio was still a rough and tumble sort of place in 1884.

That year the San Antonio Electric Company was operating two additional dynamos, one supplying twenty arc lights and the other supplying twenty incandescent lights. The dynamos were belted to an automatic Westinghouse steam engine of sixty horsepower, supplied by an eighty-horsepower boiler. By 1888 the company was providing electricity for several hundred incandescent lamps, had successfully lit Main, Military, and Alamo Plazas, and had built a four-mile distribution system for supplying electricity to businesses and residences. Businessman and promoter L. S. Berg began operating the Berg Light & Power Company the same year, and that company succeeded in implementing fully operational streetlights. Berg quickly cashed in on his company's success, selling out to San Antonio Electric two and a half years later for $20,000. San Antonio Electric moved from its location on West Commerce Street to the newly acquired Berg plant at the corner of Villita and Presa Streets.

In 1889 San Antonio Edison was chartered to establish an electric street railway system, with plans to replace the existing transportation system, which used cars pulled by mules and horses, by 1890 when the city's first large power generating station would be completed. The Alamo Electric

Vaudeville owner Jack Harris was shot by gunman Ben Thompson (pictured) in 1882. Two years later Thompson was killed inside the Vaudeville.

The San Antonio Edison Company was chartered in 1889 to establish an electric streetcar system and replace mule-drawn cars.

Street Railway Company began to convert the mule-drawn streetcars into electrified trolleys by removing the car body from the old rail truck and mounting it on an electric power truck outfitted with a control mechanism. Manufacturer Julian Sprague attached a motor to each of the two axles, spring-mounted to the frame to soften the bumps. The cars connected to overhead electric wires by a pole mounted on top of the streetcar. At the end of the pole was a wheel called a trolley that ran along the overhead wire, powering the truck as the streetcar drove. The Hot Wells Line opened in the summer, and its electrified trolley went along Presa Street to the fairgrounds, south of Mission Concepción.

As always, improved technology created challenges for these early companies. More efficient alternating current began to replace direct current, but it was expensive. San Antonio Edison, which used only direct current, began to lose customers because electric lights in homes and businesses would dim whenever the streetcars passed by. San Antonio Electric's conversion to alternating current proved so costly that the company sold out to San Antonio Gas, and the new entity became the San Antonio Gas & Electric Company. More competition came along in 1896, when the Mutual Electric Light Company was chartered and began to generate at 405 East Houston Street. Since there were no franchises, all three electric companies strung their wires with abandon and began erecting a forest of poles.

As the decade ended, the remarkable changes that catalyzed the city's building boom were described in the hyperbolic writing style of the times by a New York newspaperman who attended a convention in San Antonio in 1896—after an absence of twenty-five years.

The magic wand of civilization has touched the city and "presto, change!" has been pronounced. Where stood the old adobe huts, magnificent iron, stone and brick buildings are to be seen. Instead of the yell of the Comanche, you hear the scream of the locomotive all around the city. In place of the mesquite thicket, where the coyote held his nightly revels, you see fine, broad avenues lined on either side with beautiful and stately residences, surrounded with magnificent groves of shade trees and lovely gardens of flowers. Tall spires, piercing the skies, mark the places where pious people assemble

In the late 1800s Main Plaza was a hub of activity, and newly paved streets accommodated various modes of transportation.

to worship God. The little old rickety wooden bridges have disappeared, and in their stead a hundred magnificent iron bridges now span the beautiful stream . . . the cobblestone sidewalks have been crushed and filled until they are just as smooth as the shell streets of Galveston. Many manufacturing establishments have been inaugurated. They now have a population of nearly 40,000 with three daily newspapers.

The transformation was only beginning. The next decade would bring more mergers of the city's earliest utilities, resulting in the establishment of San Antonio Public Service Company, the predecessor of CPS Energy. The influx of residents, business, materials, and capital would change San Antonio in a myriad of ways, and all of the city's stakeholders would need power to succeed.

By the end of the nineteenth century San Antonio was described as a "forest of poles," as several small independent electric companies strung their wires with abandon.

PART 3

A New Century Begins
San Antonio, 1900–1929

In the first ten years of the twentieth century San Antonio's population nearly doubled, to 96,614. The frontier town of fifty years before had disappeared, evidenced by the elaborate Sunset Station, where the Southern Pacific Railroad delivered a steady flow of enterprising men and women ready to start a life in a town on the rise. Literacy and culture were valued; a stately new public library and fourth newspaper were proof of that. Plans were under way to create a zoo, without cages, where bison, African lions, deer, and monkeys could roam on 35 acres within the beautiful

343-acre park deeded to the City by Brackenridge in 1899. That year the wealthy philanthropist and president of the San Antonio Water Works became the first president of the San Antonio School Board and began making major contributions to education.

In 1902 beer maker Otto Koehler left the city's largest brewery, Lone Star Bottling Works, to run the competing operation at Pearl Brewery. He built a large home in the Laurel Heights suburb and purchased the Hot Wells Spa and

As the twentieth century began, the Southern Pacific Railroad brought passengers to the beautiful new Sunset Station in San Antonio. Opposite: The Pearl Brewery, c. 1900.

The Hot Wells Spa and Bathhouse drew celebrities from around the country, and a special railroad spur was built to accommodate them.

Bathhouse on the south side of town. The resort's healing waters drew celebrities from all over the country, and a special rail spur brought them directly to the spa in their private railcars.

Power made this rapid growth possible. There were five hundred gas meters in operation, about half of which were used for lighting and half for measuring the flow of gas used for fuel. Thanks to the railroads, coal had made gas production cheaper and more efficient, and San Antonio Gas & Electric operated two generating plants—Station A at the intersection of South Presa and Villita Streets, and Station B on Mission Road.

The company had also become one of the biggest retailers of gas stoves, selling these "indispensable household items" at cost, recognizing that their presence in homes would increase gas consumption. A city ordinance required that all vehicles, from wagons to automobiles, be equipped with lamps for safety, and incandescent lights were the order of the day, with more than 2,200 customers lighting their homes with these marvels. The century began brightly, both figuratively and literally.

Competition among the numerous companies providing utilities grew fiercer, and their leaders recognized that consolidations were the best way forward. In 1900 San Antonio Gas & Electric absorbed the five-year-old Mutual Electric Light Company and San Antonio Edison merged with Alamo Electric Street Railway to form a single transport entity, the San Antonio Traction Company. To celebrate, San Antonio Traction employees were invited to "take their lady friends" on a ride over the company's streetcar line "in a nicely decorated and illuminated car, winding up at San Pedro Springs for refreshments."

The power industry was experiencing the same rapid growth, with the formation of small independent companies all over the country. In New York there were more than thirty companies generating and distributing electricity, although the New York Edison Company (still part of Consolidated Gas) was taking the lead. Edison's

personal secretary was Samuel Insull, an ambitious young man who had emigrated from London in 1881 to work for the inventor. As a switchboard operator at Edison Telephone Company in London, he overheard conversations about the job opening in the United States and seized the opportunity. In 1889 Insull helped his mentor establish Edison General Electric and became its vice president. When financier J. P. Morgan combined his Thomson-Houston Electric Company with Edison General Electric three years later, forming General Electric, Insull welcomed the new capital. Others on Edison's early team felt that Insull had been disloyal. Edison forgave him, and Insull moved to Chicago as president of Chicago Edison, which would eventually become the utilities giant Commonwealth Edison, thanks to various mergers and acquisitions. Edison's

Station A, built at the corner of Presa and Villita Streets in 1909, replaced the gas plant on San Pedro Creek.

New York company would eventually become Consolidated Edison, one of the country's largest investor-owned energy companies, traded since 1824 on the New York Stock Exchange under several corporate names, making it the longest traded stock in the United States.

In San Antonio, businessman W. H. Weiss, who had been the president and treasurer of the San Antonio Street Railway Company until the City sued him for recovery of taxes in 1896, initiated a slick scheme just before merger fever struck the utilities industry. He bought the majority of common stock of most of the utility companies in town, including San Antonio Gas, San Antonio Street Railway, San Antonio Edison, and Mutual Electric. After the companies completed their mergers, he petitioned for and received an ordinance from the City that set unprecedented standards, including a clause that protected the new franchises until 1940 and spelled out the general rates the companies would charge. After the ordinance passed, Weiss sold his common stock in the companies to Emerson McMillin, which transferred ownership to Southern Light & Traction Company, a subsidiary of American Light & Traction. One day after the acquisition of the utilities, the Texas attorney

general filed suit seeking to forfeit the new consolidates' charters on the grounds that merging them violated the state's antitrust laws. A shareholders' agreement was reached, and both San Antonio Gas & Electric and San Antonio Traction reappeared as incorporated entities. But this was not the end of antitrust issues for these companies, as they would appear again, much more dramatically, about thirty years later.

While New York Edison was taking the lead in generating and distributing electricity in New York, its founder was building a team of "Edison stars" to expand his network.

When Theodore Roosevelt returned to San Antonio in 1905 for a reunion of his Rough Riders, he noted that the city had been greatly modernized since the group trained there in 1898.

The paving of a few downtown streets began to cause a gradual disappearance of horses and mule-drawn carriages. Paving affected the street-cars too, since they were denied access when some streets, like Commerce, proved too narrow for passage. When Theodore Roosevelt came to San Antonio for a reunion of his Rough Riders in 1905, he rode into town on his horse, "Little Texas," and observed that the town had changed since his visit seven years before.

As San Antonio grew so did its need for electricity and fuel. The Houston Street gas plant was pioneering the use of coke gas enriched with oil and supplying it to customers. But even with enriching it proved no more efficient than coal gas, as its heat value was no greater. The plant's gas storage was not large enough to keep pace with the growing demand for electric generation, and construction of a much bigger gas holder began on Salado Street. Designed to store 500,000 cubic feet of gas, it would increase San Antonio Gas & Electric's capacity fivefold. Construction of electricity-generating units was also under way at a fast pace, and the Mutual Electric unit at Villita and Presa Streets was upgraded in both capacity and technology.

Experts in this ever more complex industry were needed, especially engineers. In 1906 William B. Tuttle arrived in San Antonio to work as general manager at San Antonio Traction. He was thirty-two years old and had the work experience the company needed. With a degree in mechanical engineering from the University of Virginia, he went to work at the Consolidated Gas Company of New Jersey, a subsidiary of American Light & Traction. He started out mixing concrete with a shovel but soon became the superintendent's assistant. The following year he moved to the holding company's corporate office

By 1917 Station A had become a modern power plant, upgraded in both capacity and technology.

in New York and was soon in charge of accounting for the companies controlled by American Light & Traction. Over the next few years he was sent to subsidiaries around the country to report on their operations, including plants in Long Island and Binghamton, New York; Quebec, Canada; Lincoln, Nebraska; Denver, Colorado; Montgomery, Alabama; and finally, San Antonio. Assigned to report on the condition of San Antonio Gas in 1906, Tuttle stayed for three months and recommended that American Light & Traction build a new gas plant. His return to New York was brief, for in 1907 he was sent back to San Antonio to manage San Antonio Traction and oversee construction of the gas plant he had recommended. For the next four decades Tuttle would hold diverse leadership roles as San Antonio's power industry developed. When San Antonio Traction merged with San Antonio Gas & Electric a few years later, Tuttle was among the leaders at the newly formed San Antonio Public Service Company, serving as vice president, general manager, and president over the years.

In Chicago, Samuel Insull's utility empire had grown in size, but as the decade began Chicago Edison was struggling. When Insull spent the Christmas holidays in his native England, he noticed that although the shops in the tourist town of Brighton were closed, their lights shone brightly. He contacted the head of the town's electric company, who shared a "secret formula" for computing billing based on low-demand and high-demand electric use instead of the standard flat rate. After developing his own rate formula for Chicago, Insull rolled out the new plan and succeeded in turning the company around. Homeowners saw their bills lowered by more than 30 percent, and Chicago Edison became profitable, pleasing Insull's mentor, Thomas Edison.

Col. William B. Tuttle was employed by American Light & Traction when he came to San Antonio in 1907 to oversee several of the company's subsidiaries.

Over the years utilities would move back to flat rates, but today "time of use" rates are still discussed and are quite controversial.

In 1907 Insull merged Chicago Edison with Commonwealth Electric Light & Power and began to convert from Edison's original direct current to alternating current. As president of the new Commonwealth Edison, Insull built the company over the next decade to 6,000 employees serving 500,000 customers. It had the largest generating stations in the country and used 2 million tons of coal annually. Insull began purchasing portions of the bigger utility infrastructure, including

Federal Signal Corporation, Northern Indiana Public Service Company, Peoples Gas, shares of many smaller utilities, electric railroads, and radio broadcasting companies.

Others in the industry watched the spectacular rise of Commonwealth Edison with amazement and wondered if Insull's confidence that state and federal regulators would recognize utilities as natural monopolies and allow them to grow with little competition would prove to be correct.

Utilities everywhere were changing fast, and jobs in the burgeoning industry offered entry level positions with extraordinary potential. Just as American Light & Traction's powerful chairman, Emerson McMillin, had started in the gas business by wielding a shovel in Ohio, young men willing to work hard found opportunity in this new world. A gangly, six-foot-tall nineteen-year-old named Ben Utz came to work at San Antonio Gas & Electric in 1911 as a "helper" and stayed for

In 1916 surveyors mapped out sites for expansion of Station B, later called the Mission Road Power Plant.

By 1917 mule-pulled gas trucks were being replaced by motorized vehicles.

forty-six years, eventually attaining the position of general foreman before his retirement in 1957.

"I went to work as a Helper in the Electric Department in 1911, and I worked out of the storeroom in Station A. It was a job to get the mules and wagon from the 10th Street stable each morning before work and return them every evening," Utz shared with the company's in-house magazine in 1982. "I started work for the San Antonio Gas & Electric Company for a dollar and a quarter a day. When I made Lineman in 1913, I was paid three dollars a day—big money in those days!" As the horse and buggy gave way to autos, mule-pulled wagons were phased out and electric trolleys were eventually replaced by buses. It was Utz's crew who took down the overhead wires in the late 1920s and 1930s.

Meanwhile overhead power lines for growing San Antonio neighborhoods, now stretching into rural areas, were being constructed at a fast pace, and the men who installed them needed both skill and courage. "In those days, lineman training was a matter of pulling yourself up by your own bootstraps," Utz explained. "The neophyte lineman had to learn on his own. I'd practice at lunch time. The first pole I climbed was at Comal and Colorado Streets. One time I fell off a pole behind the old Hot Wells Hotel when a crossarm broke. In my day, with one hundred pound crossarms to lift, and very little equipment except blocks and tackle, the work was hard. It wasn't until 1942 that equipment started picking up; and when I retired in 1957, there were two bucket trucks that we used to get up to the insulators on the highlines. I predicted that within ten years, linemen would be using buckets."

Construction was booming in San Antonio. The military presence was building, with Air

Early work crews strung lines and climbed poles without safety equipment; skill and courage were essential to the job.

More than 250,000 men were organized into aero squadrons by soldiers at Kelly Field as the United States prepared to enter World War I.

Corps personnel training under way at the Kelly and Randolph air fields. With the Mexican Revolution raging a few hundred miles to the south and World War I brewing in Europe, San Antonio was considered a strategic military position for the United States. When a German submarine torpedoed the *Lusitania* in 1915, Americans were among the passengers. Texas joined other states in urging Washington to sever diplomatic relations between the United States and Germany, and the Texas Preparedness Program began. Two years later the infamous Zimmermann Telegram—outlining Germany's offer to help Mexico regain territories in Texas, New Mexico, and Arizona if Mexico joined the Central Powers—was intercepted. President Woodrow Wilson asked Congress to declare war on Germany. Electrical systems, plumbing, streets,

and machine shops were installed at Kelly that summer in a thirty-day timeframe. Soldiers at Kelly organized more than 250,000 men into aero squadrons over the next year, and a total of 326 squadrons were formed there during the war.

SAPSCo expanded its Mission Road plant and added generators and chimneys, aware that staying ahead of demand was imperative to meet growing power needs created in part by the war. The company purchased two smaller electric generating companies, and gas holders capable of holding 2 million cubic feet of gas were added to the system. By the end of 1917 manufacturing capacity peaked at 7.5 million cubic feet a day.

Citizens joined others across the country in planting war gardens, buying Liberty bonds, and participating in a food conservation program

New generators and chimneys were added to SAPSCo's Mission Road plant in 1917.

called Hooverizing, in which Monday and Wednesday meals were "wheatless," Tuesdays were "meatless," Thursdays and Saturdays were "pork-less," and sugar was to be conserved every day.

Katherine Stinson, San Antonio's renowned aviatrix, began flying her single-engine Curtiss Stinson-Special at fundraising demonstrations for the Red Cross, and on May 23, 1918, at twenty years old, she carried the U.S. mail from Chicago to New York, setting a nonstop flight record and traveling the distance at an average of seventy-one miles an hour.

More than 200,000 Texans joined the armed forces, and more than 5,000 lost their lives. It was a brutal, short-lived conflict for Americans. Following the armistice on November 11, 1918, American soldiers—many of them young boys— returned home as heroes, went back to their

Twenty-year-old Katherine Stinson flew her single-engine plane for the Red Cross in 1918, traveling at seventy-one miles an hour.

When it outgrew its headquarters in 1921, SAPSCo moved into a larger building on St. Mary's Street along the San Antonio River.

civilian jobs, and tried to recapture the sense of normalcy they had known before the war.

After World War I ended, one of the most explosive decades of the century began. The Roaring Twenties was a time of change and prosperity. Remembered for its speakeasies, flappers, Prohibition, avant-garde music, social change, automobile craze, and advancements in radio and film, the 1920s turned the nation upside down. Just as women's hemlines rose from the ankle to the knee, equally exciting changes were happening in the world of power transmission. With better distribution, more and better motors, new controls, and trans-Atlantic telephone service, the industrial age was humming.

The decade ushered in major technological advances as well, including Bell Telephone Laboratories' mechanism for recording sound electronically, Westinghouse's de-ion circuit breaker, Seimens' development of the expansion-type circuit breaker, Martin Hochstadter's three-core power cable, Luigi Emanueli's oil-filled cables, and E. B. Wedmore and W. A. Whitney's air-blast circuit breaker.

The 1921 discovery of the Eldorado oil field in Kansas brought about still another remarkable development in the electrical industry—its newfound use for pumping oil. E. L. Staley, an electrical engineer at the time, wrote: "Five years ago, if anyone had dared to suggest that more

SAPSCo gas inspectors used modern radio test cars in the early 1920s.

than half of the oil produced in the Eldorado field would be pumped out of the ground with motors, some of his friends would have a started a subscription to buy him a one-way ticket to Osawatomie." Electricity had become the expected power source for lighting a home or running a fan or sewing machine, but the idea that it could power an entire oil rig was unbelievable. "Nobody could ever convince him that a man shoveling coal under a boiler located thirty miles away from the rig could possibly produce enough power to lift a string of ten-inch casings," Staley continued. In the end, of course, the skeptics were wrong.

The military's needs for cotton and wool for tents and uniforms, leather, food supplies,

and petroleum—all of which were produced in Texas—had caused booms in those industries as well. The postwar demand for tungsten lamps, electrical refrigerators and ranges, and other electrical domestic appliances opened up a new world for electrical workers. SAPSCo had outgrown its original headquarters at Houston and St. Mary's Streets, and in 1921 the company moved to a larger building at 201 North St. Mary's.

Disaster struck San Antonio in September 1921 when the remnants of a hurricane in the Gulf of Mexico brought torrential rains that flooded the streets with as much as twelve feet of water, causing fifty-one deaths locally and more than $5 million in damages. Olmos Creek

reached its peak in just one hour, and homes near Brackenridge Park were inundated with as much as ten feet of water. The *San Antonio Light* described the "wall of water that carried houses from their foundations, swept motor cars away, destroyed concrete bridges, tore down trees and poles and ripped up paving in the streets like . . . pebbles."

The company's gas system manager, W. D. Burk, described how the water topped the walls of the Station A plant on Villita Street, prompting workers to throw the switches to shut off the city's electricity. The boiler fires at the Market Street waterworks were drowned by floodwaters, the telephone system went down, and telegraph lines were out. Floodwaters reached the second floor of SAPSCo's new headquarters. The city was cut off from the rest of the world. Two days later electricity was restored to most of the unflooded areas,

The flood of 1921 devastated San Antonio and disabled the delivery of water and power to the city for a few days. SAPSCo crews worked double shifts to restore electricity.

A bridge destroyed by the flood Sept 10th 1921 [people killed]

SAPSCo played a major role in the restoration work and in planning efforts to avoid another disastrous flood.

telephone service worked in about one-fourth of the city, and streetcar conductors were driving their own automobiles, transporting passengers along the streetcar lines that still had no power.

The huge job of restoration and repair began, and SAPSCo played a major role in the efforts. Utz recalled working for fifteen days straight on repairs to damaged poles and lines, snatching short naps whenever possible, and calling home on the rare occasion he could get to a telephone that worked.

City commissioners established a Flood Prevention Committee to assess how San Antonio might avoid a similar disaster in the future.

Attorney Harry Rogers, banker Franz Groos, and department store owner Nat Washer, along with engineers W. B. Tuttle, Edwin Arneson, Clinton Kearney, and Willard Simpson were appointed to serve.

Residents had diverse views on what should be done, but all agreed that a dam was needed on Olmos Creek. For the next several years engineering surveys were made and at last, in 1924, citizens approved a $2.8 million bond issue to pay for construction of a dam.

Lewis Fisher's book *American Venice: The Epic Story of San Antonio's River* describes the flood and the building of Olmos dam just north

In 1922 the Mission Road plant underwent more modernization, overseen by a team of electrical engineers known as the "Electric Stars."

of town, at a cost of $1.5 million, in superb detail. Samuel Crecelius, a retired colonel in the U.S. Army Corps of Engineers based in San Antonio, was chosen to design the project. He had worked on two dams near Laredo and had directed dam projects in several other states. After two years of construction, with some design changes and the difficulties that often occur with big projects, the dam was completed in 1926. Fisher describes it as a "concrete wonder 80 feet high and 1,925 feet long, with a 1,100-acre retaining basin, where a park and golf course were planned." Some of the engineers recommended adding a paved sewer, but the San Antonio Conservation Society protested and defeated that idea.

Not long after the flood, E. F. Kifer, SAPSCo's vice president and general manager, hosted a tour of the new Mission Road plant for the city's Rotary Club. "Our big problem is to find business for our large investment in equipment during off peak hours, or during the valley of our load," he told club members. The challenge of power distribution and generation is an age-old problem in the industry, as important today as it was a century ago.

That fall Kifer established an in-house publication for the company's employees. He envisioned the magazine as a "publication by and for the employees of the San Antonio Public Service

Company," recognizing that building community and team spirit would create an optimal experience for employees, and ultimately for their customers. A variety of sports and social clubs were also organized, including a bowling league and a baseball team. The *Broadcaster* brought employees all the news of their company, as well as interesting facts about the city and country they lived in. An early issue reported that it cost nearly twice as much to supply the nation with cigars and cigarettes as it did to serve 9 million customers with gas for cooking, heating, and lighting, noting that "the nation's gas bill in 1921 was $411,195,503, while $527,259,000 was spent on candy and ice cream."

The discovery of natural gas in South Texas near Three Rivers later that year changed the region's energy world forever. In a front-page story in the December 22 *Broadcaster*, editor Sid Ballinger broke the exciting news to SAPSCo employees:

> December 22 will long be remembered in San Antonio. It registers a change from the use of manufactured gas to natural gas; it represents the realization of a long-cherished dream of local gas consumers, and it means that the Christmas turkey will sizzle and roast over the flames of fuel unleashed from the earth and carried in pipes for more than 60 miles. . . . Santa Claus slipped down the municipal chimney a few days ahead of schedule to deposit his wonderful gift in long waiting stockings.

Preparations began at SAPSCo to convert power plants from manufactured gas to natural gas, a challenging process since natural gas was dry and the types of gas previously used were "wet."

E. F. Kifer became general manager of SAPSCo in 1923.

But the changes to the system were made during the holiday season in record time and completed before the end of the year, and Southern Gas Company constructed the pipeline for delivery.

In January 1923 Tuttle was promoted to president, succeeding New Yorker Alanson P. Lathrop. Like Kifer, Tuttle believed in communicating often with the employees, recognized how hard they had worked to complete the conversion, and considered them partners in a great energy adventure, a philosophy he ingrained in the organization's culture over the years. As the year began, he published this message in the company magazine:

A contract was signed that year with the American Car Company to purchase ten streetcars. They were lightweight—about 28,000 pounds each—and carried forty-eight passengers. The streetcar system peaked at ninety miles of track in 1926, but as new communities like Terrell Hills and Olmos Park were developed in the hilly areas farther away from the city center, travel in

streetcars was slow going, sometimes hampered still more when young pranksters greased the tracks on the Monte Vista route. The public grew increasingly dissatisfied with that mode of transportation, and passengers began to complain.

SAPSCo had been experimenting with the concept of bus service since 1921. It built its own vehicles in a local shop and ran a service to and from Fort Sam Houston, which never allowed streetcars onto its vast military facility. The reaction to buses was positive, and in 1923 the company acquired its first factory-built buses, using them for outlying areas not served by its streetcars.

In 1926 construction began on a power plant on the banks of the Comal River outside Landa Park in New Braunfels, thirty miles northeast of San Antonio. It was built to use lignite coal, which came from deposits in central Texas, for power generation. By the end of 1929 the Comal plant handled practically all of the electrical load for both cities. It continued to burn lignite until 1930, when it was converted to natural gas.

Even the new plant could not meet the city's growing power needs, and SAPSCo purchased two smaller generating units from local owners. More substations were added and transmission lines crisscrossed the city, except for downtown, which had opted for underground lines. The company's engineers developed new techniques and methods of generation and distribution to keep pace with state-of-the-art electrical service.

In 1929 San Antonio's industrial gas and electricity business increased 51 percent over the previous year. The San Antonio Express Publishing Company replaced its water-powered presses with electricity-powered ones, and the

In 1921 SAPSCo manufactured its first buses in its garage in response to customers' desire for motorized transportation.

Milam Building became the country's first fully air-conditioned office building and the St. Anthony its first fully air-conditioned hotel. The Smith-Young Tower, with six elevators, and the Majestic Theatre, with its sound system and lighting, were just a few of the other businesses that increased their power demands. SAPSCo had also contracted to supply power to a cement plant under construction and to Randolph Field. A $2 million, 30,000-kilowatt generating unit was installed at the Comal plant to help meet the city's growing electricity needs.

All previous gas consumption records were broken that year, and the natural gas industry was expanding rapidly. New gas pipelines were under construction, including one that stretched from Texas to Mexico, crossing rivers, mountains, and swamps. Modernized pipelines were constructed

The Comal plant was dedicated in 1927, and by 1929 it handled almost all of San Antonio's electrical load.

with welded joints and steel, coated to resist erosion. Feeder lines snaked into all parts of the city and its suburbs, promising to deliver better service even in severe cold weather conditions.

SAPSCo's fleet of vehicles had grown by the late 1920s, and its garage on Jones Avenue was expanded.

Spray pond technology was used at the Mission Road plant, facilitating the heat transfer process to the plant's generators.

When a December ice storm in northwest Texas destroyed the transmission lines run by the Texas Power & Light Company and the West Texas Public Utilities Company, SAPSCo came to that area's assistance, operating the Comal plant at its highest capacity ever. Houston also needed help from SAPSCo when it lost power due to mechanical problems at the Houston Lighting & Power Company. This was the start of a long tradition of power restoration support that continues today.

A spray pond was built at the Mission Road plant to cool the hot injection water before its reuse. A predecessor to cooling tower technology, the spray pond facilitated the heat transfer process in the 1920s and 1930s; a few are still in use today.

It seemed that nothing could stop the march into a prosperous new decade. The city's first skyscraper, built in 1928, was the country's tallest brick and reinforced concrete structure and its first high-rise air-conditioned office building. With twenty-one stories, the Milam Building towered over the skyline at 280 feet. Designed by George Willis, a locally acclaimed architect, it was named for Ben Milam, a hero of the Texas Revolution.

Another building, designed to be even taller, was already under construction a few blocks away. Its developers, brothers Albert and Jim Smith, envisioned a complex of buildings where a former section of the San Antonio River had been filled in and readied for construction. The brothers were two of eleven siblings who grew up in Crockett, Texas. In 1904 they went into business together running a livery stable, working on family farms, dabbling in real estate, and building some of the state's first paved highways. With their profits they established Smith Brothers and developed roads and real estate projects in several states. In 1927 the company built several buildings in San Antonio, including one at the corner of Navarro and Villita Streets (present-day CPS Energy headquarters), the Plaza Hotel (later Granada Homes), and the Federal Reserve Building (later the Mexican Consulate). The star of their projects was the Smith-Young Tower, named for themselves and their attorney, J. W. "Jim" Young, a longtime friend from Crockett who served as vice president of Smith Brothers Properties.

Designed by renowned architecture firm Ayres & Ayres and built by McKenzie Construction Company, the eight-sided neogothic brick and terracotta building towered more than 400 feet, with thirty-one floors, six elevators, and the new luxury of air-conditioning. When it opened in June 1929, more than 5,000 people came to celebrate, riding the elevators and exploring the tunnel beneath St. Mary's Street that led to the Plaza Hotel. There were speeches, food and drink, and dancing on the roof past midnight.

The stock market crash in October 1929 caught businessmen like the Smith brothers, and the nation and the world, by surprise. Anyone who was spread thinly over too many projects found himself with more money going out than coming in. Huge, seemingly impregnable companies like U.S. Steel and General Electric tumbled, and major financial centers like New York and Chicago collapsed. Rumors of investors jumping out of buildings

In 1929 citizens celebrated the opening of the ultramodern Smith-Young Tower, which had thirty-one stories and was equipped with six elevators.

spread through Wall Street, and the country watched in horror as life savings were lost in an instant. Despite the holiday lights in the streets a few months later, there was an uneasy knowledge that dark economic times were ahead. San Antonio and its businesses geared up for the fallout to come.

Unemployment lines were long during the Great Depression, and citizens gathered at City Hall in 1930.

PART 4

Rollercoaster Ride
San Antonio and Beyond, 1930–1942

Before Samuel Insull's holding company empire collapsed in the Wall Street Crash of 1929, it was valued at more than $1 billion and was comprised of railroads, new radio stations, and many utilities, including Commonwealth Edison in Chicago. Insull had graced the cover of *Time* magazine twice, in 1926 and 1929. But with only $27 million in equity, the company was doomed when the stock market failed, and more than 600,000 investors lost their life savings. Insull and his wife fled to France and then to Greece to avoid federal prosecution for antitrust violations. Harold Ickes, a newspaper reporter and political reformer with a law degree, vowed to bring him to justice.

Ickes grew up in Chicago, was class president of his high school, and worked his way through the University of Chicago. He earned his law degree in 1907 but never practiced. Instead he worked as a reporter for the *Chicago Tribune* and focused on uncovering corruption in his city, taking on powerful figures that included Insull, Mayor William Hale Thompson, and even Robert R. McCormick, owner of the *Chicago Tribune*. As

Samuel Insull's giant holding company was world-famous before the Wall Street Crash of 1929.

the decade progressed, Ickes's dedication to political reform would earn him the nickname "Honest Harold" and a place in President Roosevelt's cabinet. The Public Utility Holding Company Act of 1935 is a major part of his legacy.

As the Insulls hid in Europe, and Ickes dreamed of ways to crack down on large utility holding companies, San Antonio felt the first fallout from the Great Depression. In a letter to his employees, SAPSCo's General Manager Kifer warned workers to avoid borrowing money from the many "loan sharks" who had appeared on the scene, advising that "this company will not tolerate in the future any of its employees making a business of loaning money to other employees." He offered help in the form of advance wages, in cases of dire emergency, and stepped up extracurricular activities and sports to build camaraderie.

SAPSCo's baseball team was especially successful that year, thanks to a young part-time employee and ringer named Jerome "Dizzy" Dean, who would pitch two winning games for the St. Louis Cardinals in the World Series a few years later. A health column began to appear in the *Broadcaster*, written by Dr. C. O. Sappington, with advice for avoiding the common cold and tips for general well-being during hard times. Sappington suggested that "the one thing that keeps more men from losing their balance when everything seems to have gone awry is an interesting and absorbing hobby." The company ramped up its social support system and filled the newsletter's pages with happy announcements of marriages and babies, sports, and stories that reassured employees that normalcy coexisted with the stress of the times.

A worrisome development for cities like San Antonio with large military bases was news of downsizing the armed forces throughout the

Political reformer Harold Ickes vowed to seek justice for the victims of antitrust violations, and when he became secretary of the interior he pushed hard for reform.

Jerome "Dizzy" Dean, a part-time employee at SAPSCo in the early 1930s, eventually played for the St. Louis Cardinals in the World Series.

country. Fortunately city leaders managed to thwart base closures and secured a new Air Corps training facility by donating 2,300 acres of land valued at $500,000. This project generated jobs and an increased demand for power. Of course, before the company addressed how that power would be supplied, critical infrastructure would be needed to connect to the power source. Southwest Dairy Products built a new headquarters and, to showcase its striking architecture, installed expensive 1,000-watt floodlights provided by SAPSCo's new lighting department. Power from the Comal plant supplemented the two in-town plants on Villita Street and Mission Road.

Advances in design and materials supported capabilities for generators to go from lower to higher pressures, providing more efficiency and more energy output. The Mission Road plant installed two of the new high-pressure units—with modern wheels and dials to monitor them—displaying the remarkable evolution that had taken place since the belt-driven dynamos delivered power to cities a few decades before.

By 1930 the longest pipeline in the world stretched from the West Texas gas fields to northern California. Every three hundred miles along the route, compressor stations with cooling towers kept gas flowing safely, allowing California's Pacific Gas & Electric Company to provide electricity to the citizens of San Francisco and neighboring towns. The same year three holding companies, including Insull & Sons, began a $60 million project to build a pipeline from northern Texas to Chicago using a technique for coating the pipeline directly in the field. The other partners in the newly incorporated Natural Gas Pipeline Company of America were Standard Oil of New Jersey and City Service Company, owned by Henry Doherty, an oilman and financier from

Advances in design and materials enabled generators to go to higher pressures, increasing efficiency and energy output.

Ohio. The new line would allow Insull to convert his Peoples Gas Light & Coke Company's service area from manufactured coal gas to cleaner, hotter, cheaper natural gas. Once the pipeline began to deliver natural gas in fall 1931, Chicago became the first major city to completely convert its utility distribution system.

Around the same time, another holding company consortium purchased a pipeline project from Houston oilmen William C. Moody and Odie R. Seagraves, who had completed only a small portion of the line designed to transport gas from Houston to Omaha, Nebraska. Partners in Moody-Seagraves Interests were North American Light & Power, Lone Star Gas, and United Light & Power (which owned American Light & Traction). They called the entity the Northern National Gas Company and built a 1,110-mile pipeline that eventually transported gas from Texas to Minnesota via Omaha.

Natural gas for SAPSCo was supplied by the Magnolia Gas Company via a twenty-inch pipeline to the huge Bruni Field in South Texas. Magnolia later merged with Southern Gas to form

El Paso Natural Gas Company, established in 1929, built pipelines sixteen inches in diameter to deliver natural gas from West Texas to northern Mexico, Arizona, and eventually California.

the United Gas Pipeline Company, which supplied the city with natural gas for electric generation and distribution until 1961. SAPSCo, Pacific Gas & Electric, and Peoples Gas Light & Coke were all subsidiaries of conglomerates that operated on an increasingly complicated and intertwined national stage. By 1932 the country's eight largest utility holding companies controlled 73 percent of the country's investor-owned electric industry, and Washington was worried.

Closer to home, construction of Randolph Field was making great progress, "enclosing within its boundaries the population and conveniences of a small town," according to news articles in summer 1931. All power—approximately 150,000 kilowatt-hours a month—was supplied by SAPSCo, and the water tank on top of the facility's tower held 500,000 gallons of water.

H. B. Zachry, a young engineer from Laredo, Texas, had started the H.B. Zachry Company a decade before, after graduating from Texas A&M University and working for a few years at the Texas Highway Department. His first major project was the construction of four concrete bridges in Laredo, and his business grew rapidly in South Texas in the mid-1920s. Zachry was awarded the contract for concrete paving at the huge military facility, his first job of many in San Antonio; eventually his projects would span the globe.

With the influx of personnel at Randolph Field, more public transportation was needed, and the utility company purchased six new twenty-one-passenger yellow buses. They were ultramodern, with six-cylinder, ninety-six-horse-power engines, hydraulic and vacuum-booster brakes, balloon tires, and individual leather seats.

The construction of Randolph Field in 1931 was a massive undertaking, with immense positive impact on the city during the Depression years.

Once these luxurious vehicles were added to the fleet, the company was operating ninety-seven buses and 160 streetcars.

Modernization was under way inside the company's offices as well. Telephones were installed, marking the end of the switchboard era. There were two models with dials, and illustrated instructions for using these marvels were distributed to all employees. Utilizing radio technology to improve its communication with customers, the company sponsored a weekly show on the city's first radio station, WOAI, presenting San Antonio's history from its early days.

Despite the prosperity of the utility company, the Depression's impact was everywhere. As the holiday season approached, the cover story of the *Broadcaster* delivered a clear message:

> Merry Christmas. That's all. This year we should enter into the spirit of the season with renewed vigor and interest. Thousands in our community are unemployed and many are underfed and underclothed. It is a challenge to those more fortunate. If you really want a "Merry Christmas" such as you have never had before, enlist today in the army that is rushing to the relief of San Antonio's

poor and distressed. Give, and give until it hurts. . . . Be thankful that you have a job and can give. A thousand times better employed and giving than unemployed and begging.

The worst times hit San Antonio a little later than the rest of the country, and as 1932 began citizens were keenly aware of it.

Longtime fears of higher wage scales and unionized labor had resulted in strong resistance to heavy industry, causing the city commissioners to turn down overtures by Ford Motor Company to build a new plant in 1925. Now, as unemployment lines grew longer, there were questions about the wisdom of that decision. The city's history of reliance on light manufacturing, tourism, and the financing and processing of agricultural and ranching products from South Texas had molded an economic vision that was beginning to unravel. The value of total building permits declined from $18 million in 1929 to half that amount in 1930. In an attempt to bolster the economy and create jobs, voters approved $5 million in public works bonds, but even this did little to lift the gloom. By 1932 building permits had dropped to 10 percent of the previous year. The fanfare of 1929, when citizens danced with abandon on the rooftop of the dazzling Smith-Young Tower, seemed a distant memory.

The *Broadcaster* again reached out to SAPSCo employees with this advice:

Start the new year determined to be a good representative of the company that has continued to give you a living wage and splendid working conditions while millions have been turned into the streets. Employees of this company are indeed

fortunate, and those who do not realize it and appreciate it should call for their time slips and move on.

The crisis had forced one-third of the city's twenty-one financial institutions to close, and the failure of City Central Bank & Trust Company shook San Antonio to its core. Tuttle traveled to Washington, D.C., to lobby for financial help for San Antonio. After negotiations with the Reconstruction Finance Corporation, he succeeded in securing a commitment for a $1.5 million loan for South Texas Bank & Trust. This was an important victory for the bank's depositors, and Tuttle urged them to join him in bringing "a golden smile of relief to the entire city." Seven hundred of the institution's largest shareholders were able to liquidate bad assets, secure the loan, and reemerge as South Texas Bank & Trust.

Unemployment was between 20 and 25 percent, however, and homeless men with families in tow became a daily reminder of hard times. The Red Cross and Salvation Army offices overflowed, and shantytowns sprang up along the banks of the river. Poverty and homelessness put a strain on city government, and many working citizens could

Modernization at SAPSCo was under way, evidenced by new telephones and a department dedicated to customer service and troubleshooting.

During the Great Depression a growing number of unemployed, homeless citizens built camps along the San Antonio River.

not pay their property taxes. City commissioners, led by Mayor C. M. Chambers, took drastic measures, eliminating all city-funded relief efforts, cutting wages in half, turning off streetlights, and terminating more than three hundred city employees.

Despite the desperate times, American Light & Traction and most of the companies it held were surviving, even doing well. As the year ended, the *Wall Street Journal* announced that it was "recommending utility stocks for investment opportunities, due to impressive earnings in the industry despite depressed business conditions."

Two officials from American Light & Traction visited their San Antonio–based utilities company in 1932. Vice President Chester N. Chubb and Treasurer James W. Laurence met with SAPSCo's management team and were impressed by the new substation in the Beacon Hill area and the showroom in the company's building, with eleven models of refrigerators for sale. Like the gas stoves a few years before, these appliances were marvels that would change both the food and power industries. Advertisements described the refrigerators as "available in all sizes and types, attractively displayed against backgrounds of white Arctic scenes," and urged customers to come see them. Chubb and Laurence returned to New York with glowing reports of SAPSCo's operations, praising the management, and were delighted with their investment.

That year when Insull's creditors forced his holding company into bankruptcy, the press called it "the biggest business failure in the history of the world." Protests about monopolistic trends were growing louder, and pressure from investors hurt by the financial manipulations of other large holding companies increased. Some of the electric, gas, and street railway companies that had been negatively affected by their holding companies' irresponsible pyramiding began to complain as well. With Franklin D. Roosevelt's election in November 1932, they gained a champion. The new president chose his favorite reformer and friend Harold Ickes as secretary of the interior, and the two men set their sights on the holding companies that controlled so much of the nation's electricity.

Hoping to avoid federal intervention, the utilities industry voluntarily established regulations that included a forty-hour work week and a minimum wage, and the Edison Institute predicted that the codes would result in an increase in employees from 350,000 to 400,000 over the next few years. There would also be more gradual increased expenditures for the utilities, and these would need to be passed along to customers.

The *Christian Science Monitor* weighed in on the growing debate. "The ideal public utility is not difficult to define. It is an organization of men and equipment established and maintained to render first-class service to its customers at reasonable rates based on sound financing of the enterprise, a cultural wage to its employees, and a fair return on a just capitalization." Tuttle and D. A. Powell, SAPSCo's general manager, praised the

A substation was built to serve the city's fast-growing Beacon Hill area.

publication's definition in their year-end message, and it has remained at the core of San Antonio's electric utility.

Part of the company's "first-class service" entailed listening to customers' requests. Heading the list was the desire for modern vehicles. In 1933 SAPSCo paid the City $250,000 to end its streetcar contract seven years before its expiration. On April 29 an old "mule car" led a parade of retiring streetcars from San Fernando Cathedral up Broadway to the Witte Museum, with SAPSCo staff cheering them on. Newspaper articles noted that they had been "the most important factor in the growth of the city since 1878" and that San Antonio was the first city of its size to convert to all-bus transportation.

A few relics of San Antonio's first public transportation system became part of an exhibit at the Witte Museum, and the other cars were sold in various places, including an eastern brokerage firm that purchased twenty for use on Third Avenue in New York City.

In 1933 SAPSCo retired its streetcars and replaced them with buses. The last "mule car"
led a parade of streetcars from San Fernando Cathedral to the Witte Museum.

Mayor Chambers died suddenly later in 1932, and Charles Kennon Quin was selected to serve the last few months of the term. Quin was born in Louisiana in 1877 and spent his childhood in Weimar, Texas, where his father served as mayor. He opened a law office in San Antonio in 1923 with Chambers as his partner, and in the late 1920s and early 1930s he was an assistant city attorney and a city utilities attorney.

When mayoral elections came up again, Quin was elected to a regular term. He followed the political machine tradition of the times and knew how to call in votes. He looked to longtime associate Charles Bellinger to deliver the support of the

city's African American population. In return, Quin's city government provided those neighborhoods with paved and lighted streets, plumbing, a meeting hall, a branch library, and improved schools.

Bellinger grew up in a farming family in Caldwell County and moved to San Antonio in 1906 to establish a saloon with loans from Otto Koehler and the Pearl Brewing Company. He was an exceptional gambler and entrepreneur and quickly expanded his businesses to include a café, cab company, real estate and construction company, theater, barbershop, private lending service for African Americans, lottery, and bootlegging

SAPSCo employees watched the parade signaling the end of the streetcar era. Tuttle is pictured in the front row, fourth from right.

operation during Prohibition. He became active in politics in 1918, developing support for John W. Tobin, whose grandfather was the first mayor of San Antonio in the Republic of Texas. Tobin served as sheriff for twenty-one years and was elected mayor in 1923. A decade later Bellinger delivered those same votes to Quin.

Despite an injunction filed by newly elected Texas Attorney General James Allred against the sale of electric and gas appliances by SAPSCo and others, various sales campaigns continued while the company's attorney, Wilbur Matthews, and former governor Dan Moody represented the utility companies in court. Eventually the case reached the state supreme court, which ruled in 1934 that "the sale of electric and gas appliances was authorized as an incidental corporate power related to the sale of electricity and gas."

Around this time Insull was caught in Turkey and extradited to the United States for trial. Legendary Chicago lawyer Floyd Thompson defended the former utilities magnate, and to everyone's surprise he was found not guilty on all counts. More outraged investors joined the growing body of citizens opposing what they perceived to be powerful utility companies. President Roosevelt began pushing hard for federal legislation that would facilitate their regulation, proposing that they limit their operations to a single state where they would be subjected to effective state codes, or by forcing divestures so that each would become a single system serving a limited region.

In his State of the Union address in January, Roosevelt delivered a strong message to citizens of the United States, promising to work toward "the abolition of the evil of holding companies." The next month House Speaker Samuel Rayburn of Texas and Sen. Burton K. Wheeler of Montana introduced the Public Utility Holding Company Act on the floors of the House of Representatives and the Senate. Debates raged and dishonest attempts to influence legislators were uncovered over the next few months, during what historians describe as "the fiercest congressional battle in history." Howard C. Hopson, president of the New York–based holding company Associated Gas & Electric, organized a campaign to send hundreds of telegrams opposing the bill to legislators. They came from people who did not exist and were paid for in cash. When the scheme was uncovered, Hopson was investigated for his role and for paying himself and his family more than $3 million between 1929 and 1933—equivalent to approximately $45 million in today's dollars. To make matters worse, the company's 350,000 investors did not receive a single dividend. Before his rise to the top of Associated Gas & Electric, Hopson earned his CPA degree, was admitted to the bar, and worked for the Interstate Commerce Department in Washington, D.C., and for the New York Public Service Commission. In 1915 he established a consulting firm in New York whose clients included Associated Gas & Electric, AT&T, Consolidated Edison, and Electric Bond & Share Company. Seven years later he purchased Associated Gas & Electric for $298,318.19 with

Howard Hopson (left), president of the Associated Gas & Electric Company, with his attorney, Willam Hill, in 1935.

money he borrowed from his sisters. By 1930 his holding company owned more than 250 companies providing utilities in many states, the Philippines, and the Canadian Maritime provinces. According to an article in the December 1935 issue of *Fortune* magazine, Hopson was said to have boasted that "the only laws he could not get around were the ones he himself wrote." When his stockholders sued him, however, he was found guilty of defrauding them of $20 million and was sentenced to prison.

After various amendments and compromises, the Public Utility Holding Company Act was passed and signed into law on August 26, 1935. The large holding companies hoped they could thwart its enforcement, and many filed lawsuits against the Securities and Exchange Commission. They also mounted media campaigns to bolster support from the press and their customers, who were beginning to believe that the Great Depression was over.

San Antonio's economy showed enough improvement to reconnect 350 of the more than 1,500 streetlights that had been dark for the past few years due to cuts to the city lighting budget. A confidence was in the air, and citizens began to purchase more than the bare necessities. SAPSCo's Trial Range Program in South Texas to boost sales of gas ranges and the Better Light—Better Sight drive to sell lamps to would-be customers were both highly successful. By the end of 1935, nearly 25,000 lamps had been sold and the company began to advertise a free commercial lighting service to local businesses.

SAPSCo hosted the Southwest's first air-conditioning show the next summer, displaying five models. As a result, eighteen units were installed in homes, theaters purchased twelve, and restaurants bought nine. Fans were still the most popular cooling device, and 750 were purchased in June alone with an easy down payment of ninety-five cents. But San Antonio was fast becoming a national center for air-cooling machines, and several companies, including Friedrich, began to move from refrigeration into air-conditioning. The industry had come a long way since Willis Carrier, a young graduate of Cornell University, invented the electric air conditioner in 1902. The Nix Hospital became the country's first air-conditioned hospital, and the Milam Building was the country's first completely air-conditioned office building.

Holding companies and their subsidiary utility companies were keenly aware of potential changes once the Public Utility Holding Company Act took real effect. In San Antonio Powell remained in his dual roles of vice president and general manager, and longtime president Tuttle was made chairman of the company's board.

On August 26, 1935, President Franklin Roosevelt signed the Public Utility Holding Company Act with its authors and supporters at his side.

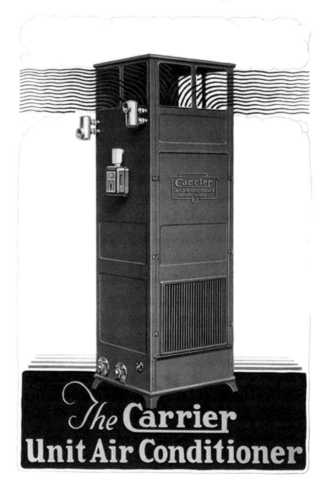

In 1936 SAPSCo hosted the Southwest's first air-conditioning show, displaying five different models.

Chester N. Chubb, formerly executive vice president of American Light & Traction, relocated from New York to become president of SAPSCo. Chubb had been a frequent visitor to San Antonio, brought years of experience as a utility company executive, and was well liked by the company's local management team. He was also American Light & Traction's man on the ground as the big holding company kept a careful eye on its subsidiaries during changing times. In his year-end letter, Powell referred to "the past year of uncertainties" and urged employees to maintain a united front and work for "increased business, more friends, and an adequate solution to our present problem." Teamwork was declared the slogan for 1936.

Nationally syndicated journalist Walter Lippmann weighed in with a column in the *Daily Oklahoman* titled "The Utility Mess—And How to Get Out of It." He criticized the antagonism between the federal government and the utilities as being "a little short of being a disgrace to the country" and urged all parties to consider compromise. Lippmann reminded readers that the electric industry was still a "new industry" in the hands of bold and enterprising men who were pioneers, sometimes "courageous to the point of recklessness." He outlined the industry's remarkable development over a short time, praising its ability to supply electricity at diminishing costs and increased efficiency and acknowledging occasional abuses and "flagrant profiteering." Lippmann urged Roosevelt to soften his stance, noting that "a reasonable man need not be afraid or ashamed to make peace. For though he may be taunted by the irreconcilables, the good will of reasonable men will be a sufficient compensation."

Further complicating issues was the Tennessee Valley Authority (TVA), which was flexing its

Following passage of the Public Utility Holding Company Act, D. A. Powell (pictured) retained his dual roles as vice president and general manager of SAPSCo and Chester Chubb, executive vice president at American Light & Traction, relocated from New York to become president of SAPSCo.

muscle in ways that worried private utility companies operating in the region. Organized in 1933 as a cornerstone of Roosevelt's New Deal, the TVA was charged with providing navigation, flood control, electricity generation, and economic development to one of the country's most depressed areas. Largely within Tennessee, the region included parts of Kentucky, Georgia, Mississippi, and North Carolina, where the average annual income was $630, crop failure was rampant, and 30 percent of the population suffered from malaria.

The concept of government-owned generation facilities selling to publicly owned distribution systems was controversial and remains so today. In 1935 the TVA proposed selling surplus power at the Wilson Dam. Fourteen preferred stockholders of the Alabama Power Company filed suit, recognizing that this would pose a threat to their market. The case went all the way to the Supreme Court, which upheld the TVA's right to sell surplus power at government-built dams and "to build transmission lines to transport such power to a reasonable market." Over the next few years the construction of dams in that region displaced more than 15,000 families, and many of those shared the anti-TVA sentiments of the private utility companies. Others welcomed the improvements the federal agency brought to the struggling region, including James Agee, a young journalist who wrote in *Fortune* magazine that "in this enormous machine, the balance is human."

Similar projects were under way in Texas. The state legislature created the Lower Colorado River Authority in 1934 to complete a partially built dam that one of Insull's companies had abandoned in 1931 after the holding company collapsed. State Senator Wirtz was its champion, and in Washington U.S. Congressman James P. Buchanan worked hard to secure funding from the Works Progress Administration. The following year Wirtz introduced legislation to create a similar organization—the Guadalupe-Blanco River Authority. Both were modeled after the TVA, designed to control flooding of the Colorado and Guadalupe Rivers and to develop, conserve, and protect water resources. A fact that was unspoken but recognized by astute entrepreneurs and politicians was that they also had the potential to impact the utilities industry in powerful ways.

A young graduate of Southwest Texas State Teachers College in San Marcos, Texas, had been watching these developments from a desk in Washington, and he saw an opportunity. Lyndon Baines Johnson had grown up in the Texas hill

The Tennessee Valley Authority's construction of Wilson Dam fueled controversy over government-owned generation facilities selling power to publicly held distribution systems.

country, in a house with no indoor plumbing or electricity. He struggled in school but found his calling when newly elected U.S. Congressman Richard Kleberg hired him as an executive assistant in 1931. In Washington, Johnson pored over every detail of protocol and threw his energy into Roosevelt's New Deal programs and into making political connections at the national and state levels. He became an ardent protégé of Buchanan and Rayburn, making them feel like father figures, and their strong recommendations to the president won Johnson a new position as Texas state director of the National Youth Association, established in 1935. He dispersed federal funds around the state, promoted employment opportunities for the young people of Texas, and continued building a remarkable political machine.

Buchanan Dam in the Texas hill country was dedicated in 1937. Lyndon B. Johnson and Alvin J. Wirtz envisioned building a power program in Texas like the Tennessee Valley Authority's.

Two years later Buchanan died from a heart attack, after serving in Congress for twenty-four years. Johnson immediately asked Senator Wirtz for help seeking the vacant congressional seat. Wirtz's support meant access to political contributions and connections to influential people. The canny politician also provided a strategic idea that some historians believe won Johnson the special election and his "miracle seat" in Congress. Wirtz told the twenty-eight-year-old to align himself with Roosevelt's sweeping social policies and to campaign the hardest in Texas's smaller cities, which the other candidates ignored. Johnson was elected and returned to Washington with his wife of three years, Claudia "Lady Bird" Johnson, determined to make a difference in the battle against poverty. An early goal was to bring electric power to the Texas hill country of his youth.

The following year, when the Lower Colorado River Authority began operating the dam that Insull's companies had started, they renamed it Buchanan Dam in honor of its staunch supporter. Johnson and Wirtz were on hand for the dedication in late 1937. As proponents of the TVA, both politicians envisioned building a similar public power program in Texas.

Meanwhile there was growing skepticism of the TVA's reported successes. In its monthly bulletin to members, the Edison Institute provided this analysis of the TVA's 1938 income statement: "To have 'broken even' on its own power allocation, TVA's rates should have been raised by 80%; to have constituted a yardstick, still on its own allocation, TVA's rates should have more than doubled; to have constituted a true yardstick on

By 1937 there were 51,000 gas customers in San Antonio, gas rates had been reduced to less than half the 1920 rates, and 2,000 modern meters had been installed over twelve months.

a proper allocation of construction expenditures, TVA rates should have more than tripled."

The debate about public versus private ownership was just beginning and would last for more than a decade. Former Virginia governor and U.S. Senator Harry F. Byrd weighed in early, asserting that "the pump of private enterprise can only be run by the motor of confidence, and that motor will run only when the private business man is assured that he will get a fair deal from his government." He represented the conservative sentiments of Roosevelt's predecessor, Herbert Hoover, who denounced big government and opposed federal intervention in business affairs; others in

Congress felt the same way. This encouraged the large holding companies as they prepared to fight the legislation calling for their divesture. Wendell L. Wilkie, president of the Commonwealth & Southern Corporation, pointed out "that in the decades of the 1920s, there was nine dollars of private corporate financing for every dollar of government financing, and in the 1930s, there was only thirty cents of private financing for every government dollar." But the strain of the Great Depression had also fueled the opposing view that more government regulation and assistance could pull a struggling population through a terrible time. The debate was far from over and would eventually find new voices in the heated discussions about hydroelectric power that continue today.

Not yet impacted by this debate, or by competition from the TVA or any other entity, SAPSCo employed 1,492 people and experienced steady growth in 1936. The transit department carried 18,400,000 passengers and owned 225 buses. The gas department laid eleven miles of gas mains, installed 1,200 services, set 2,000 meters, and had 51,000 customers. Gas rates had been reduced to less than half the 1920 rates. The electric department had constructed 120 miles of poles, installed 2,100 services, set 4,500 meters, and had 71,000 customers. Its rates were reduced by 12 percent from the previous year. The South Texas department added six towns to its grid—Old D'Hanis, Castroville, Utopia, Dunlay, Zorn, and Bracken—bringing the conveniences of power to rural customers. Frank Brothers Department Store installed air-conditioning to the delight of its customers, the Santa Rosa Hospital added two more air-conditioned operating rooms, and more than 450,000 passengers rode city buses to Fiesta Week events in 1937.

The Works Progress Administration updated the municipal airport at Stinson Field, adding lighted runways that enabled planes to land after dark.

That year the Works Progress Administration, established in 1935 by Congress, completed construction of an updated municipal airport at Stinson Field, adding a modern administration building and longer, lighted runways where young Katherine Stinson had once taught would-be aviators to fly planes traveling at seventy miles an hour.

As the decade drew to a close, distribution consultant Elmer Roper delivered the keynote address at the Edison Electric Institute's convention in New York. His talk, "What the Customer Thinks of Utilities," focused on customers' three most important concerns—perception of prices and services, courtesy and "neighborly qualities,"

and community involvement. Tuttle, always one of the city's most active civic volunteers, deeply believed in all three concepts and had made them an essential part of SAPSCo's culture. Company leaders who followed him carried the tradition into the future.

The city was not without its scandals, however. On December 30, 1938, with an election looming, Mayor Quin was indicted for misappropriation of funds. A Bexar County grand jury charged him and two other city officials with allegedly using city funds to pay a day's wage to more than four hundred "precinct workers" in the previous year's primary election.

SAPSCo headquarters on St. Mary's Street, c. 1939.

Two years earlier Quin's old friend Bellinger had been convicted of failure to pay income taxes and was sent to prison at Leavenworth penitentiary. Quin and other city leaders petitioned President Roosevelt, and Bellinger was granted parole. Quin's charges were eventually dropped as well, but his opponent, Maury Maverick, won the election. Maverick served only one term, and Quin became San Antonio's mayor again in 1941.

Modernization was under way in SAPSCo's three power plants, and the company was increasing the number of substations it operated to meet the needs of a growing population. While the Comal plant in New Braunfels expanded, the gas generating plant on Salado Street, where Tuttle first worked, was being dismantled. Its biggest tank had been utilized as a relief holder for manufactured gas; now it would be used for natural gas emergency storage.

As the end of the decade approached, the Works Progress Administration conducted a survey in fifty-nine cities to assess how the average American family spent its annual income of $1,250. Results showed that 35.6 percent went to food, 17.6 percent to shelter, and 14.6 percent to clothing. The remaining 30 percent was spent in a variety of smaller categories, with electricity claiming just 1.5 percent.

As the 1940s began, the Mission Road plant was updated to meet needs of the growing population and in anticipation of the possibility of U.S. involvement in World War II.

As the country pulled out of the Great Depression, 1939 brought an increase of 12 percent in the electric industry. In its report to the stockholders, SAPSCo reported a 7.5 percent increase, explained partly by the fact that the country's more industrialized regions were taking the lead in electricity usage. But the company's rates were still lower than the national average, and it was keeping pace in appliance sales. Military presence in the city was building again, fueled by the possibility of another world war in Europe, the economic rollercoaster ride of the 1930s appeared to be over, and the new decade looked promising.

In May 1940 company president Chester Chubb published a letter to employees in the *Broadcaster*, assuring them that American Light & Traction "has no desire to sell its holdings in SAPSCo, which it regards as one of its most valued holdings." He detailed some of the Public Utility Holding Company Act's uncertainties, explaining that American Light & Traction's parent company, United Light & Power, had filed papers with the Securities and Exchange Commission "reserving the right to contest the validity of the Act, as well as the jurisdiction and power of the SEC to require it to dispose of its investment in any other company." He concluded with the strong message that "AL&T's holdings are not for sale."

In the same issue, Chubb relayed a request received from Fort Sam Houston's District Recruiting Office to include the following in all company publications: "We favor adequate preparedness for National Defense, and recommend enlistment in the United States Army to eligible young men." Texas National Guard officers asked for employers' cooperation to release men for summer camps and training, and Chubb complied, urging employees to coordinate absences with

their department heads so that "the operations of the Company can be carried on to the best advantage" while they were away. As World War II escalated in Europe, the nation's efforts turned to the needs of national defense, and SAPSCo recognized that the electric industry would be critical. "We have reason to be proud," Chubb announced. "Our industry is prepared for any emergency that might arise."

Meanwhile construction of a garage on the site of the Salado Street gas plant was completed under the supervision of longtime employee W. D. Burk. It boasted all sorts of modern improvements, including a welding shop and a building where gas pipes could be coated and stored before they were installed.

As the presidential election of 1940 approached, the company's community-oriented culture was again evident as Chubb reminded employees that "it is our job to serve all people of all creeds and of all religions; and politics and religions don't mix with the rendition of public service . . . our customers are Democrats, Republicans, and Independents. The public is served by giving prompt, courteous, and friendly service and allowing the customer to solve his political and religious problems in his own way."

Roosevelt was reelected, soundly defeating conservative opponent Wendell Wilkie, who had been the great hope for holding companies facing the legal decisions that the Public Utility Holding Company Act would bring. The year concluded with the knowledge that a war in Europe was raging and that the past decade had not brought the peace and joy people hoped for. But despite increased expenses at SAPSCo, the normal growth of San Antonio—along with emergency growth due to the defense program, and a very

SAPSCo constructed a new garage in 1940 to accommodate a growing number of company vehicles.

cold winter—boosted the company's successful year-end performance. Its report to stockholders and employees outlined plans for new boilers and electric generating units at several plants, as well as the need to order twenty-five buses for transportation and scheduled improvements to the gas distribution system and its pipeline capacity.

By now employees of SAPSCo and the rest of the nation were asking if the United States was going to war, and many employees were already in military training camps. Patriotic columns and reports from the front filled the pages of the *Broadcaster* and most of the country's newspapers and magazines.

When the Japanese bombed Pearl Harbor on December 7, 1941, the question was clearly answered. Company Christmas parties were canceled, and moneys intended for holiday celebrations were diverted to America's "victory drive" with the realization that many SAPSCo employees would be going off to war.

Early in 1942 war bond programs were set up, and SAPSCo encouraged employees to buy them for as little as $2, deducted from their paychecks. "Remember Pearl Harbor!" was the resounding slogan that year. Demands on utility service in all departments showed big increases. The jump in public transportation passengers was especially large with the shift from civilian automobile

IN THE NEWS

FOR CLASSIFIED CALL F-1231
International News Dispatches Appear Exclusively in The Light

THE SAN ANTONIO LIGHT
AN INDEPENDENT TEXAS NEWSPAPER
Member of the Associated Press ★ A Constructive Force in the Community

HOME EDITION

VOL. LXI—NO. 323.   Published by The Light Publishing Company, San Antonio, Texas.   MONDAY, DECEMBER 8, 1941.   TWENTY-TWO PAGES   THREE CENTS Per copy in the city and vicinity. Five cents on trains and elsewhere

U. S. DECLARES WAR

3000 Killed, Wounded in Hawaii Raids; U.S. Ships Sink Nippon Submarines

(Map on Page 6-A; Pictures on Page 5-A.)

By the Associated Press.

The White House acknowledged today a bloody toll of 3000 killed and wounded in the Japanese attack on Honolulu—about half of them fatalities—as the battle of Hawaii continued and imperial Tokyo headquarters claimed smashing naval and air victories over the United States.

Great Britian and Australia formally declared war on Japan, an American war declaration was drafted for swift passage by congress, and mighty forces of the U. S. fleet were reported combing the waters of the Pacific to seek battle with Japanese warships.

The White House said several Japanese submarines and planes had been accounted for and active resistance was "still continuing" against the Japanese attacking force in the vicinity of Hawaii.

Reinforcements of planes are being rushed to the islands, the White House said.

Bombers Fly From Mainland.

Meanwhile, Tokyo newspapers carrying unofficial identification of the two U. S. battleships purportedly sunk said they were the 29,000-ton Oklahoma, built in 1914, and the 31,800-ton West Virginia.

A Domei broadcast asserted 60 per cent of the United States entire naval power was stationed in Hawaiian waters prior to the attack and the surviving units "would be regarded as utterly inadequate to accomplish any successful outcome in an encounter with the thus far intact Japanese fleet."

The White House said an old American battleship turned over in Pearl Harbor and one destroyer was "blown up."

Bombers flown from San Francisco were said to have arrived in Hawaii while the battle was raging.

In the Far East, a British communique said Japanese air raiders killed 63 persons and wounded 133 today in a violent assault on Singapore, Britain's "Gibraltar of the Orient," but Japanese troops were being "mopped up" in an attempted land invasion of Malaya from the north.

An Italian broadcast quoted Domei as listing the 33,100-ton U. S. S. Pennsylvania and the Oklahoma as the American battleships sunk. Two United States destroyers and two oil tankers were also reported lost.

Japs Claim Forces Unscathed.

The Tokyo announcement asserted there were no Japanese losses in striking the heavy blows against the United States fleet at Honolulu.

While Americans waited for some word from Washington of United States counter-blows, the Japanese reported 50 or 60 U. S. planes had been shot down in air combats over Clark field, in the Philippines, and another 40 over Iba, 80 miles north of Manila.

Only two Japanese planes were acknowledged lost.

Only two Japanese planes were acknowledged lost.

The Japanese also announced an agreement between Japan and Thailand for transit of Japanese troops through Thailand—presumably for an attack on British Malaya, site of Britain's great Far East fortress of Singapore, or British Burma. Both adjoin Thailand.

Japanese troops were reported to have landed at two points on the Gulf of Siam, far down the Thai coast near Malaya.

An official British announcement at Singapore said Japanese warcraft which landed troops at two places in British Malaya, near the Thailand frontier, had been put to flight.

Japanese forces still remaining on the beach were being heavily machine-gunned, the British said.

Domei, the Japanese news agency, was quoted as saying Japanese and British troops already were fighting in Thailand.

A Reuters dispatch said it was announced officially in Washington that Thailand had ceased resisting a Japanese invasion army temporarily and that negotiations were under way.

In Manila, Admiral Thomas C.

Hart, commander of the U. S. Asiatic fleet, announced a small contingent of American marines at Peiping, China, had been forced to surrender to overwhelming Japanese forces.

An N. B. C. broadcast said the U. S. aircraft carri'r Langley was reported unofficially in Manila to have been damaged in action with Japanese planes.

A C. B. S. broadcast reported at least 250 casualties inflicted by high-flying Japanese planes in two attacks on the Philippines.

Manila itself apparently had escaped attack thus far.

A U. S. army bulletin said 30 Japanese bombers attacked Davao on Mindanao island and bombed Baguio, the summer-time capital of the Philippines. One Japanese plane was reported shot down in Davao bay.

Francis B. Sayre, American high commissioner to the Philippines, declared the situation was "well in hand."

A WOR-Mutual broadcast from Manila, repeating Japanese parachute troops had landed on the islands, said native Japanese had

(Continued on Page 6, Col. 5)

Men Flock to Colors

S. A. Goes on War Footing

San Antonio was on a full time war footing Monday.

News of the attack on United States Pacific possessions by Japan electrified the city Sunday afternoon.

Lights burned all night in army headquarters.

Monday morning additional guards paced the military reservations.

(Continued on Page 6 Column 4.)

Kelly Flier Killed in Jap Action

By the Associated Press

The following is the list of members of United States armed forces killed in the war in the East, as disclosed by official advices to the next of kin:

First Lieut. Hans Christiansen, 21, Woodland, Calif., Marine aviator, at Pearl Harbor.

Pvt. George G. Leslie, 20, Arnold, Pa., army air corps, at Hawaii.

Robert Niedzwiecki, 22, Grand Rapids, Mich., at Hawaii.

Lieut. James Derthick, 22, Ravenna, Ohio, army air corps at Honolulu.

Second Lieut. Forge A. Whiteman, Sedalia, Mo., air corps, at Pearl Harbor. (Trained at Randolph and Kelly field, Texas.)

Gordon Mitchell, Kensington, Kan., air corps, at Hawaii.

Pvt. Donald Plant, 22, of Wausau, Wis., air corps, at Wheeler field, Hawaii.

Pvt. Dean W. Cebert, of Gatesburg, Ill., at Honolulu.

Japs Raid Near Manila

MANILA, Dec. 8.—(AP)—Japanese bombers struck at military bases and ports the length of the Philippines today, smashing at the big Fort Stotsenburg, Clark field, the summer mountain capital at Baguio, the port of Davao and Aparri and the far northern Batan island group.

Manila had heard no air raid alarms and seen no raiding planes early tonight almost Japanese warcraft were reported within 40 miles of the densely populated city.

Manila, which has no public air raid shelters, was blacked out beneath heavily overcast skies from short after dusk. Other ports and shut off lights and waited tensely.

The army headquarters announced Japanese population on the southern Davao, center of concentrated Japanese population in the southern Philippines, and Aparri and Baguio, summertime mountain capital of the Philippines north of Manila, had been bombed by daylight.

During the afternoon Japanese bombers struck at Fort Stotsenburg, one of the biggest army encampments in the Philippines, and nearby Clark field.

Numerous buildings were said to have been set afire and the army's telephone communications to Manila were cut.

Private advices from fort Stotsen

(Continued on Page 6, Col. 5.)

Japs Rounded Up; Canal Guarded

BALBOA, Canal Zone, Dec. 8.—(AP)—With the United States maintaining a wartime guard over the vital Panama Canal Zone, the Panama government decided to intern all Japanese residents on the isthmus and affirmed its intention to co-operate fully with the United States.

The roundup by Panama or Japanese aliens, began soon after word was received yesterday of the Japanese attack on American outposts in the Pacific, proceeded smoothly during the night while the U. S. army rushed construction of an internment tent city.

Panama police reported 130 of the 300 Japanese residents in the city of Panama had been taken in

War on Japs Declared By Britain

LONDON, Dec. 8.—(AP)—Britain, like the United States under Japanese attack, declared war today on the Tokyo government, without waiting for Washington first to formulate an American declaration.

Said Prime Minister Churchill:

"It only remains now for the two great democracies to face their tasks with whatever strength God may give them."

At the same time Britain made allies of Thailand and Free China.

Prime Minister Churchill told the house of commons Britain had been forwarded to the British embassy at Tokyo and that at 1 p. m. (6 a. m. C. S. T.) a note was handed to the Japanese charge d'affaires here "stating that in view of Japan's wanton acts of unprovoked aggression the British government formed them that a state of war existed between the two countries."

Churchill recalled that "with the full approval of the nation and of the empire I pledged the word of Great Britain about a month ago that should the United States be involved in war with Japan, a British declaration would follow within the hour."

Churchill declared Britain had assured Thailand "that an attack on her will be regarded as an attack on us" and that he had messaged Generalissimo Chiang Kai-Shek of Free China, saying that that henceforward we would face a common foe together.

NO NEED TO WAIT

Churchill disclosed he consulted President Roosevelt in a Transatlantic telephone call last night "with a view to arranging the timing of our respective declarations." He explained the president informed him only the congress could declare U. S. at war.

"I then answered him we would follow immediately. However, it soon appeared that British territory in Malaya had also been the object of Japanese attack and later on it was announced from Tokyo that the Japanese high command—a curious form—but the Imperial Japanese government but the Japanese high command—had declared that a state of war existed between them and Great Britain and the United States."

That situation, Churchill went on, left no need to wait for a congressional declaration as in Washington.

"The house stood and cheered.

He spoke of recent reinforce

(Continued on Page 6, Col. 3)

S. A. Clouds Wiping Out Rising Sun

Cold, damp and murky weather was the menu for San Antonio Monday and Tuesday.

With light to moderate northerly winds blowing, high Monday was to be about 62 and Tuesday about 54.

Considerable cloudiness was to bring occasional light rain.

High Sunday was 61 and low Monday morning, 49.

Cuba Congress Asked for War

HAVANA, Cuba, Dec. 8.—(AP)—The cabinet asked Cuba's congress today to declare war on Japan.

TEGUCIGALPA, Honduras, Dec. 8.—(AP)—Honduras declared war on Japan today and the government established martial law throughout the republic.

PORT-AU-PRINCE, Haiti, Dec. 8.—(AP)—Haiti joined the Latin American nations today which have declared war on Japan and pledged the United States his full assistance.

Parents Told Son Killed by Attack

WOODLAND, Calif., Dec. 8.—(AP)—Mr. and Mrs. Peter Christiansen were notified here today by the navy department that their First Lieut. Hans Christiansen, 21, marine aviator, had been killed in action at Pearl Harbor which bore the brunt of the Japanese attack yesterday.

Belgian Diplomats Will Leave Tokyo

LONDON, Dec. 8.—(AP)—The Belgian government in exile instructed its ambassador in Tokyo today to leave Japan with British and American diplomats, (P. R. O. in New York heard a B. B. C. announcement that the Belgian refugee government definitely had decided to break with Japan.)

Japs Take Wake Island, Say Nazis

BERLIN, Dec. 8.—(INS)—The German wireless reported tonight the Japanese have occupied Wake island, disrupting air connections between the United States and the Far East, despite denials from elsewhere. The denials said the U. S. gunboat Wake had merely been taken over at Shanghai.

Need for Men in U.S. Navy Critical

NEW YORK, Dec. 8.—(INS)—The navy announced today that "the need for men is critical" and that it was recruiting officers in the country would remain open 24 hours a day and seven days a week. The announcement said the age limits for enlistments are 17 and 50.

U.S., Jap Ships, Planes Battle

HONOLULU, Dec. 8.—(INS)—Warships and planes of America, Britain and Australia were locked in far-flung naval battles with Japanese forces, and bombs today over the sudden turn of events in the Pacific.

Honolulu emerged from a total blackout this morning. U. S. army and navy planes filling the skies from which Japanese had swooped in their "suicide" attack.

Lone Vote Cast Against Conflict When Congress Answers F.D.R. Request

WASHINGTON, Dec. 8.—(AP)—The United States, through its congress, declared war today on Japan.

The senate vote of 82 to 0 and the house vote of 388 to 1 told their own story of unity in the face of common danger. The speed with which the two chambers granted President Roosevelt's request for a declaration of war was unprecedented.

The single adverse house vote was that of Miss Jeanette Rankin, Democratic congresswoman from Montana, who was among the few who voted against the 1917 declaration of war on Germany.

The officially-announced toll of two warships and 3000 men dead and wounded in Japan's devil toll on Hawaii was fresh in the minds of the legislators.

The senate and house had assembled together to hear the declaration. They cheered him enthusiastically and then pushed the resolution through with not a moment's waste of time.

"I ask," the chief executive declared, "that the congress declare that since the unprovoked and dastardly attack by Japan on Sunday, December 7, a state of war has existed between the United States and the Japanese empire."

The president said that yesterday—"was a date which will live in infamy."

TEXT OF MESSAGE

The text of Roosevelt's war message:

"To the congress of the United States:

"Yesterday, December 7, 1941—a date which will live in infamy—the United States of America was suddenly and deliberately attacked by naval and air forces of the empire of Japan.

"The United States was at peace with that nation and, at the solicitation of Japan, was still in conversation with its government and its emperor looking toward the maintenance of peace in the Pacific.

"Indeed, one hour after Japanese air squadrons had commenced bombing in Oahu, the Japanese ambassador to the United States and his colleague delivered to the secretary of state a formal reply to a recent American message. While this reply stated that it seemed useless to continue the existing diplomatic negotiations, it contained no threat or hint of war or armed attack.

DELIBERATE ATTACK.

"It will be recorded that the distance of Hawaii from Japan makes it obvious that the attack was deliberately planned many days or even weeks ago. During the intervening time, the Japanese government has deliberately sought to deceive the United States by false statements and expressions of hope for continued peace.

"The attack yesterday on the Hawaiian islands has caused severe damage to American naval and military forces. Very many American lives have been lost. In addition, American ships have been reported torpedoed on the high seas between San Francisco and Honolulu.

"Yesterday the Japanese government also launched an attack against Malaya.

"Last night Japanese forces attacked Hongkong.

"Last night Japanese forces attacked the Philippine islands.

"Last night the Japanese attacked Wake island.

"This morning the Japanese attacked Midway island.

"Japan has, therefore, undertaken

"As commander-in-chief of the army and navy I have directed that all measures be taken for our defense.

"Always will we remember the character of the onslaught against us.

"No matter how long it may take us to overcome this premeditated invasion, the American people in their righteous might will win through to absolute victory.

"I believe I interpret the will of the congress and of the people when I assert that we will not only defend ourselves to the uttermost but will make very certain that this form of treachery shall never endanger us again.

"Hostilities exist. There is no blinking at the fact that our people, our territory and our interests are in grave danger.

"With confidence in our armed forces—with the unbounding determination of our people—we will gain the inevitable triumph—so help us God.

"I ask that the congress declare that since the unprovoked and dastardly attack by Japan on Sunday, December seventh, a state of war has existed between the United States and the Japanese empire.

"FRANKLIN D. ROOSEVELT.
"The White House,
"December 8, 1941."

The text of the joint resolution declaring war on Japan follows:

"Declaring that a state of war exists between the government of Japan and the government and the people of the United States and making provisions to prosecute the same.

"REPEATED WAR ACTS."

"Whereas, the Imperial government of Japan has committed un-

(Continued on Page 6, Col. 3)

Forecast

San Antonio and vicinity: Considerable cloudiness with occasional light rain Monday night and Tuesday. Light to moderate northerly winds.

Temperatures: High Monday about 62; low Tuesday morning about 44; high Tuesday afternoon about 54; high Sunday 61; low Monday morning 49.

East Texas (east one-hundredth meridian): Fair to partly cloudy in the northern portion. Considerable cloudiness in the south portion. Occasional light rains Monday night and Tuesday in the southwest portion and near the middle and lower portions. Little change in temperature. Light to moderate, mostly northerly winds on the coast.

West Texas (west one-hundredth meridian): Considerable cloudiness

WAR MAP

The fold page war map of the Orient which The Light announced for publication in Monday's paper, is not printed due to the sudden turn of events in the Pacific.

Japanese nationals were being rounded up as police stationed.

Secret defense plans for guard-

[Left column editorial]

WELI, fellow Americans, we are in the war and we have got to win it.

There may have been some difference of opinion among good Americans about getting into the war, but there is no difference about how we should come out of it.

We must come out victorious and with the largest V in the alphabet.

We are not completely prepared for war.

We have not got a Swiss system of universal service that we will have to have some day, since the lands are full of robbers and sea of pirates.

But we will get better and stronger every day, and we will not have to get very good and very strong to knock the everlasting daylights out of Japan.

We may have some small reverses at first, but do not let that worry you—if it happens.

It is not who wins the first round but who wins the last one that counts for victory.

And there is no doubt about the victory, folks—none whatever.

THE worst thing about the war with Japan is that it will divide our efforts and prevent the all out aid to England that we were doing and planning further to do.

But we will still manage to keep Britain going with our right hand while we poke Japan in the nose with the left.

Japan has been wanting war for a long time.

It has been swaggering around Asia, murdering a lot of unarmed Chinamen.

Now it is going to get a war and a real one.

Fortunately we are well on our way towards a dominating and determining two ocean navy and an all skies aeroplane fleet.

Fortunately we can manufacture 10 ships to Japan's one, and 19 aeroplanes to Japan's one.

Naturally we can fly the planes better and fight the ships better.

And that means that as soon as we swing into action we will wash up the war.

JAPAN'S attack on Hawaii is probably with the idea of keeping us on defense at home.

But we will not stay at home and we will not stay on defense.

Before the war is over we will have burned up all the paper houses in Japan and sunk most of their scrap iron battleships, put this bunch of oriental marauders back on the right little, tight little, out of sight little island where she belongs.

And we will have fenced them in there.

Then maybe we will let them have a little oil—coal oil or castor oil, we cannot tell which yet.

Our main concern now is about England.

This attack by Japan upon us is largely to create a diversion. We must not be diverted any more than is necessary for our own protection.

THE war is OUR war now—not only in Asia but in Europe.

We have got to win in both arenas.

The European war, to be frank and factual, is not going to be so easy, but we can win it and will.

We will do our best to help England now, and after we have washed up Japan we can concentrate on Europe and straighten things out there.

The politicians have had the war all to themselves for a long time.

It has been a wordy war—a windy war.

Now it is going to be a fighting war.

The American people are going to take hold.

The politicians proposed the war, and the American people are going to dispose of it.

There is going to be a new order in Europe and a new order in Asia all right, all right.

The American people are go-

As a new decade began, big changes were ahead for the world, the nation, and SAPSCo.

driving to bus riding, a result of strict gasoline and tire rationing. Some months incurred an increase as high as 42 percent.

News reached San Antonio in March that the Securities and Exchange Commission had ordered American Light & Traction to dispose of its interest in SAPSCo, with one year to comply. Chubb, as a member of the board of the parent company, promised to keep his employees informed of all developments. He assured them that their jobs were not in jeopardy but acknowledged that the future was unknown.

The race was on. SAPSCo was well run, financially successful, and available for purchase. Interested parties began to circle the prey. Two bidders for SAPSCo's common stock were soon out in front—the Guadalupe-Blanco River Authority and the City of San Antonio. The prize was a company with assets of $35 million, including gas, electricity, and transportation systems and fifty miles of electric lines extending from San Antonio in all directions. Three power plants, including a modern steam-generating plant of

60,000-kilowatt capacity, were part of the package. Both bidders had enlisted local, state, and national politicians, bankers and business leaders, lawyers, the media, and the general public in the stormy competition.

On August 30, 1942, a hurricane in the Gulf of Mexico took a sudden detour and slammed into San Antonio. SAPSCo rushed to restore power to the city. In its fury, the hurricane broke trees, tore down electric service poles and telephone poles, blasted roofs from thousands of homes, and shattered glass windows. Wires were broken, poles were smashed in two, and trees were pulled out by the roots. Damage was estimated at more than $1 million. The utility company's overhead system was destroyed, and it was estimated that 50,000 services were out at one time. Lights went out instantly in many areas. Every man in the electric department was rushed into service, and more workers were recruited from other departments.

As fall approached, America was at war, a storm had ravaged San Antonio, and SAPSCo's future was uncertain.

PART 5

City on the Rise
San Antonio, 1942–1962

The autumn was beautiful in San Antonio, with cooler weather providing some relief to the group of men—dressed in three-piece suits and ties—who had gathered in the law offices of Dewar, Robertson & Pancoast on the third floor of the National Bank of Commerce building on Main Plaza. After months of negotiations, the city attorney, a team of municipal bond attorneys from the Chicago firm Chapman & Cutler, and SAPSCo's attorneys, led by Wilbur Matthews, had hammered out a settlement between the two bidders vying for the purchase of SAPSCo. The plan called for the City to issue revenue bonds to acquire SAPSCo's stock, dissolve the company, and immediately acquire the gas and electric utilities that served the City. The City would then lease the Comal plant to the two river authorities for thirty years and buy the electricity from the Comal plant for use in the city electric system.

On October 24 the $33,950,000 bond indenture was in place, and the lease agreement with the Guadalupe-Blanco River / Lower Colorado River Authorities was signed. The establishment of the

City Public Service Board as a nonpolitical entity governed by four trustees and the sitting mayor of San Antonio guaranteed that good business practices, not city government, would guide the new company. The men in the room were confident that the deal would change San Antonio's future in remarkable ways. Mayor Quin summed it up when he announced that "revenues from these utilities will go a long way toward financing City government and growth."

American Light & Traction sold its subsidiary, SAPSCo, to the City of San Antonio on October 24, 1942.

Almost immediately after the complicated closing was completed, Matthews alerted the board that he had spotted a serious conflict. The trust indenture securing the revenue bonds required that all electric and gas system properties be held and operated by the CPSB as long as any of the thirty-year revenue bonds were outstanding. According to Matthews, the thirty-year lease agreement with the Guadalupe-Blanco River Authority was not in compliance with the trust indenture. A few days after the closing, CPSB trustees notified the GBRA that it would not surrender possession of the Comal plant, and litigation ensued. The state supreme court eventually decided the case in 1947, and the lease on the Comal plant was sustained. As disappointing as this ruling was for San Antonio, during the five-year battle the CPSB had prepared for any outcome, making improvements to its other plants and adding a power plant. Ultimately the court ruling did not have any impact on the company's ability to deliver the power its customers required.

As the five board trustees tackled these challenges, Chubb resigned, opting to continue his employment with American Light & Traction. To Tuttle's relief, a colleague from earlier days agreed to return, bringing a wealth of experience and institutional knowledge to the job. Kifer had been SAPSCo's vice president and general manager in the 1920s and early 1930s. He cheered with the team when natural gas was discovered in 1922, oversaw construction of the Comal plant in 1927, and led the company's expansion to rural areas through the formation of SAPSCo's South Texas department. Before he left in 1933 to take charge of another American Light & Traction subsidiary in Muskegon, Michigan, Kifer had contributed to an era of tremendous growth that included the interconnections of transmission systems and

Wilbur Matthews became the principal attorney for SAPSCo in 1929 and served in that capacity for the CPSB until his retirement in 1983.

The City sold SAPSCo's public transportation system to the Smith-Young Corporation. Bus tokens displayed its name—San Antonio Transit Company—and in 1943 the Smith-Young Tower was renamed the San Antonio Transit Tower.

installations of large generating units. Employees, who recognized him as someone who knew the ropes and was part of the team-oriented culture that was so important for success, enthusiastically welcomed him back.

One result of the City's purchase of the utilities company was the decision to sell its transportation department. It was severed from the gas and electric departments immediately and put under control of several local banks. W. W. Holden became president, overseeing operation of its buses while waiting for completion of its purchase by Smith Brothers and the Smith-Young Tower Corporation, now a subsidiary of Dallas Rupe & Son, for $301,100. In December the transportation system was incorporated as the San Antonio Transit Company, and the following year the Smith-Young Tower was renamed the Transit Tower.

In February 1943 the *Broadcaster* published this update about the new city-owned utility company's transition:

> The City Public Service Board is slowly unwrapping the package that was laid on its doorstep on October 25th last year. It was a big job—it is a big job—and the Board is to be commended for the progress it has made. Change from private to public ownership has brought some changes in policy because of different laws governing public enterprise. Some changes have been hard on workers, but there are many compensating factors. . . . We are now employed in an essential industry with every reason to look forward to permanent employment at wages prevailing for the kind of work we do. While it is now a municipal enterprise, the weakness

of such ventures in the past has usually been the result of political activities. We believe that the present Board proposes to steer clear of such hazards. Businessmen of the type who comprise the Board will naturally proceed with caution . . . and have proven their ability by successfully operating other businesses.

By 1943, 22 million Americans were engaged in war work and the country had produced almost 50,000 airplanes, 32,000 tanks and self-propelled artillery, and 17,000 antiaircraft guns larger than twenty millimeters. No new automobiles were in production, as steel was needed for more important purposes, and rubber tires were unavailable to the general public. By Christmas two hundred employees of City Public Service were in military service, representing more than 25 percent of the company's workforce.

The company stepped up its efforts to sell war bonds to employees, with a powerful reminder in the *Broadcaster* that "somewhere an American boy has a rendezvous with death tonight. Whether on a carrier's deck in the Pacific, amid the burning sands of North Africa, or in the tropical forests of New Guinea—whether swift and perilous, or terrible and slow his going—he is giving to his fellow men the greatest gift a man can give . . . BUY BONDS."

As 1944 began there were whispered predictions that an Allied victory might be in sight, coupled with the somber realization that thousands of American soldiers would return home disabled and that casualties had surpassed 165,000. Stories about bravery in the field and support efforts on the home front appeared in growing numbers. Newspapers announced that the Brooke Army Medical Center was readying its facilities for returning veterans, with strong backing from city leaders, including Tuttle, whose military connections had helped bring bases to San Antonio since the early 1900s.

Utility companies' role in the military effort was highlighted in news reports. According to the Federal Trade Commission the American electric industry's annual contribution to the war effort exceeded the combined production of the Axis powers. The commission predicted that "this great army of kilowatts . . . will eventually bring the totalitarian powers to their knees." In San Antonio citizens were proud that the power generated at the Comal plant came from two giant 30,000-kilowatt steam turbines with blades that traveled at 600 miles an hour. Although the validity of the lease agreement between the City and the GBRA was still being contested in court, the lawsuit did not impact delivery of electricity. The peak load on the electric system continued to be met despite increased demand from San Antonio's military bases, which now numbered five. Plans were under way to replace the low-pressure steam units at the Mission Road plant with a high-pressure 25,000-kilowatt unit, and early design plans were being considered for a plant at Leon Creek.

During World War II, 25 percent of CPSB's workforce was enlisted in the armed forces, and many trained at Randolph Field as parachutists.

To comply with municipal governance regulations, the company revamped its employee benefits. Tuttle and Kifer urged workers to take advantage of voluntary plans for life, health, accident, and retirement insurance being offered

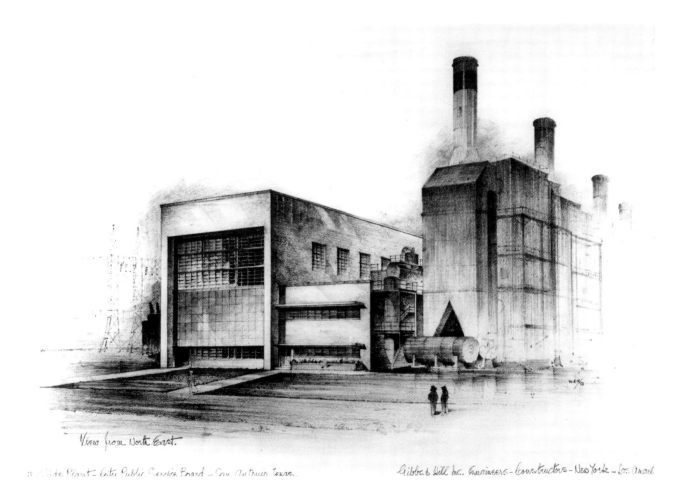

View from North East.

Plans for the Leon Creek plant were announced in 1944, and CPSB plants were upgraded with high-pressure steam turbines to meet increased energy demands during the war.

at "ridiculously low" costs. Their encouragement worked, and by summer 1944 more than 600 of the company's 725 employees had enrolled.

Tuttle and Kifer made another company-wide announcement. Effective August 1, Tuttle would resign as general manager but would remain chairman of the board, and Kifer would become general manager, a position he had held two decades before. Col. N. B. "Bernard" Gussett, who had just completed two years of service with the Eighth Service Command, was appointed executive assistant to the general manager. Prior to being called to duty in the Army in 1942, he

had served as superintendent of SAPSCo's South Texas department.

As chairman, seventy-year-old Tuttle continued to steer the Public Service Company through transition challenges and reminded employees that the board considered every customer a VIP entitled to the utmost respect, courtesy, and appreciation. He nurtured important connections he had made in San Antonio and beyond, recognizing that relationships with government officials and business leaders were important on many levels. He enjoyed friendships with some of the state's most prominent and powerful men, and as the year ended he traveled to Gov. Coke Stevenson's ranch in West Texas, along with refrigeration pioneer Richard Friedrich and the Pearl Brewery's Otto Koehler, for an antelope hunt.

In June 1944 victory on the beaches of Normandy produced jubilation in France and throughout the Allied countries. Germany surrendered in May 1945, followed by Japan three months later. The official end of World War II sparked spontaneous celebrations around the world. Within a week of V-J Day 3,000 civilians at Kelly Field resigned or retired to return to their prewar lives, some of them as trained personnel that the utility company had sorely missed. "With unbounded joy, and rampant gaiety, America celebrates the peach she has won in this horrible drama just concluded, that has affected every household that pays homage to the stars and stripes," the *Broadcaster* reported that August. "The war, and the ensuing peace, with all of their complex ramifications, are beyond conception of the human mind."

San Antonio enjoyed fast-paced growth over the next few years, and construction boomed. The $210,300 the utility provided to the City's annual operating budget contributed to that growth in myriad ways, including new outdoor lighting for the Alamo.

In 1947 gas-fired generating units were installed at the Mission Road plant, and the first 30,000-kilowatt unit was installed in the new Leon Creek plant, situated on fifty acres southwest of the city. The location was chosen because of its close proximity to the city's military and industrial complexes—including the newly established Southwest Research Institute. Oilman and philanthropist Tom Slick was just thirty-one years old when he established the nonprofit research and

Tuttle built strong business and social relationships for the CPSB, often hunting in West Texas with Gov. Coke Stevenson; Richard Friedrich, refrigeration company owner; and Otto Koehler, owner of the Pearl Brewery.

development institute, located on his 1,200-acre Essar Ranch. Six years before, he had founded the Southwest Foundation for Research and Education (after several name changes, it became the Texas Biomedical Research Institute in 2014) in a nearby complex, envisioning San Antonio as a future center for science, technology, and medical research. The laboratories and industrial plants needed power, and that year the increased capacity of the city's power plants easily met peak demands on the electric system, making the state supreme court's decision to uphold the lease of the Comal plant to GBRA less consequential.

CPSB installed outdoor lighting for the Alamo in the late 1940s.

A ring of transmission towers now encircled San Antonio more than fifty miles from the center of town, ensuring that power distribution would keep up with an expanding city. High tension wires hummed with electricity, vibrating thousands of miles in a breathless second and providing power to industry and everyday citizens. The *Broadcaster* described electricity as "miraculous, indefinable, mysterious . . . and a vital necessity to our way of life."

In April 1949 Kifer died of pneumonia after a fifteen-month leave of absence during which his executive assistant Gussett guided the company. Employees and board members mourned the loss of a leader who had witnessed the changes that natural gas brought to the energy business, expanded the company's power plants, and welcomed employees back from the war. In August Victor Braunig became the general manager. An engineer who had graduated as valedictorian from Texas A&M University in 1910, Braunig had worked for the utility for twenty-four years. Employees welcomed him warmly in his new role.

When the Leon Creek plant added a second generating unit, it became the utility's largest plant.

As the 1950s began the CPSB had 1,375 employees; the manpower shortage of the war years had ended. By 1951 the Leon Creek plant's second unit was running, making it a 66,000-kilowatt facility. In a message to customers, Braunig explained that while the plant had been designed to utilize Texas's vast quantities of natural gas, it could also burn oil in an emergency and could readily be converted to burn lignite if that became a more economical fuel source. Conveniently located railroad spurs and large lignite deposits in Milam County owned by the CPSB would make the changeover to that fuel source seamless. Ample

land had been purchased to meet future expansion needs, and there was a large supply of artesian water beneath the Leon Creek tract. Remarkably, the plant required only twenty men to run at full capacity, for twenty-four hours, seven days a week, every day of the year.

In an effort to demystify the modern power plant for civic leaders attending its dedication, Braunig described the process of producing electricity in dramatic language that did not require an engineering degree to understand. His description was later printed in a brochure that went to CPSB customers. The intensity of the process and incredible science behind it almost read like science fiction:

Energy comes from natural gas that is burned in super-heated boilers, in a seething mass of flames, with gas fires burning turbulently in a huge two-story room at temperatures of 2,200 degrees. Over 7.5 million cubic feet of gas are burned daily in each of the two identical boilers.

The sides of the boiler rooms are lined with banks of pipes through which pure distilled water rushes under high pressure. Already preheated to 350 degrees, the water is converted to steam at 900 degrees and 850 pounds of pressure in a matter of seconds. Waste products from the gases are vented off through the 140-foot smokestack. Meanwhile, the steam—under terrific pressure—is conducted through heavily insulated pipes to the turbines, traveling at more than one hundred miles an hour. It expands, causing the turbines to turn the shaft of the 93-ton generator. Powerful magnets in the component rotors whirl at 3,600 rpm across a series of coils, producing electricity from each of the two units at a pressure of 13,800 volts.

After the steam is expanded it is condensed to distilled water and put back through the system. In converting the steam to water, 60,000 gallons of cooling water per minute are used in the condensers. When passing through the condensers, the cooling water is raised from 94 degrees to 105 degrees. It is cooled to the original temperature in the giant cooling towers, then recirculated. Only a small amount that evaporates is lost in the process.

In less than a decade after the City acquired the Public Service Company, capital investment of the property more than doubled, from $33 million in 1942 to $70 million in 1951. The company was proud to report to customers that it had fulfilled all city requirements for gas and electricity, made possible by its expanded plant operations. Braunig, following the company's long tradition of year-end messages, thanked employees for their partnership, civic leaders for their support, and citizens for their confidence.

Victor Braunig, who became general manager of CPSB in 1949, oversaw tremendous expansion during his nine years in the leadership position.

Recognizing the ongoing construction boom in San Antonio, H.B. Zachry, which had done the massive concrete paving work at Randolph Field twenty years before, relocated its headquarters from Laredo in 1952. The following year it was awarded the contract to build a 66,000-kilowatt power plant on a 344-acre site that could easily handle future expansion. The location—off Perrin-Beitel Road on the far north side of town—was optimal for serving the residential and commercial areas developing to the north, and it would supplement the service supplied by the Leon Creek and Mission Road plants to military and industrial users to the south. The plant was constructed for $8,123,000, and its primary generator was taller than a ten-story building.

San Antonio, with a population of more than 400,000, was often described in promotional materials as the fastest-growing city in the United States. The San Antonio Municipal Airport had been built on 1,200 acres north of the city limits in the early 1940s. During the war it was used by the military; in 1944 it was returned to the City for civilian use and renamed San Antonio International Airport. By 1953 major runway construction and electrical installations were completed, with a passenger terminal, FAA control tower, and baggage claim area.

Modern technology brought important advancements as the decade progressed. Thanks to medical researcher Jonas Salk's discovery of a polio vaccine, there was a sigh of relief in communities around the country, as worries about children contracting the disease that killed and crippled disappeared. New Zealander Sir Edmund Hillary and his Nepalese partner, Tenzing Norgay, reached the summit of Mount Everest, the highest mountain in the world, in 1953. Walt Disney dreamed of a magical kingdom in California, and

When the Leon Creek plant was dedicated in 1951, its brochure proclaimed San Antonio the country's fastest growing major city.

when Disneyland opened in 1955 it epitomized the power of imagination. Confident that his research institutes were thriving in San Antonio, Tom Slick mounted several expeditions to the Himalayas in search of the mythic abominable snowman.

The same year appliance salesman Ray Kroc visited his best customer in San Bernardino, California. Kroc sold Multimixers, used for making milkshakes, and when he learned of a tiny restaurant that used five of them nonstop, he went to investigate. Brothers Dick and Mac McDonald

The San Antonio Municipal Airport, built in the 1940s and used by the military during World War II, was expanded and renamed the San Antonio International Airport in 1953.

had a small snack shop decorated with golden arches and a formula for providing a limited menu very quickly. Kroc convinced the brothers to give him a franchise, and the first McDonald's opened in Des Plaines, Illinois, offering hamburgers for fifteen cents. Anticipating that road systems under construction across the country would create a new sort of customer who wanted quick access to food without leaving the car, Kroc's early restaurants were situated on busy thoroughfares and offered ample parking, curbside service, and no inside seating.

In San Antonio new expressways connected neighborhoods and avoided the more crowded downtown streets. H.B. Zachry did much of the construction, and the CPSB was responsible for lighting the expressways, which by 1957 included portions of Highway 281, Interstate 10, Loop 410, and Interstate Highway 35.

On September 8, 1954, Colonel Tuttle died after a short illness. He was admired by the employees, community leaders, and government officials who had known him over many years, and his loss was keenly felt in San Antonio and beyond. Several state legislators, as well as family, friends, and colleagues from around the country, attended his memorial service. The recently completed power plant on the north side of town was renamed the Tuttle plant in his honor.

Braunig presided over the dedication. He promised guests that planning for the future was a top priority for the company and vowed to stay ahead of the city's power needs, adding that the Tuttle plant would build a 100,000-kilowatt unit in 1955 and, "assuming continued growth, another of the same capacity for completion in 1959 at a location now being studied."

Electricity sales that year increased by more than 14 percent, and gas sales by 12 percent with the installation of seventy-six miles of gas mains. United Gas had constructed a supply line that surrounded the city and fed into city gate stations, at which point the CPSB purchased natural gas to supply its customers.

The Tuttle plant's second generating unit was also 100,000 kilowatts and cost $9 million. State-of-the-art in every way, its heavy equipment came from General Electric and Westinghouse Electric, among other vendors. CPSB's management often gave tours of the facility to businessmen who marveled at the fact that only fifty years before, a 2,000-kilowatt generator had been considered huge.

Many other cities experienced similar rapid growth in their use of power, and the companies that provided it were thriving. In New York utilities giant Consolidated Edison had acquired or merged with more than a dozen companies over twenty years. Still privately owned, it represented another utility provider model and served

As modern roadways were built and the population became more mobile, McDonald's created a new restaurant model in the 1950s.

Colonel Tuttle died in 1954. The company's newest power plant was renamed in his honor the following year.

Tours of the Tuttle plant in 1955 highlighted the newest energy technology.

customers in the boroughs of Manhattan, the Bronx, Queens, and most of Westchester County. Its investors enjoyed good returns during the 1950s and 1960s and never dreamed that the future would bring an energy crisis with severe implications for the company founded by the man who revolutionized electricity.

By the late 1950s air-conditioning, with a new emphasis on modern "room coolers," helped increase electricity usage, as did the newest holiday fad—outdoor Christmas lighting. In 1956 the Empire State Building took the idea to a new level. Described in the press as "the brightest man-made light source in the world," a spectacular electric tiara crowned the famous building.

Four far-reaching night beacons, each weighing a ton and generating 2 billion candlepower, were installed at a cost of $250,000. More than 1,000 feet from the ground, they changed the Manhattan skyline and could be seen on a clear night in Boston, Baltimore, and points between.

San Antonio saw its own cityscape transformed when the CPSB decided to retire Station A, its first power plant, on Villita Street. The company made it available to the community as an assembly hall and carefully dismantled the old generators and storage tanks. Local architect O'Neil Ford consulted on the design and suggested a circular structure, free of interior columns, with a saucer-shaped roof suspended on steel cables. The unique look appealed to CPSB trustees, and they hired general contractor G. W. Mitchell to build it, requesting that much of the brick from Station A be recycled into the 29,000-square-foot structure. It opened in 1959 with ultramodern lighting and the capacity to seat 2,000 people. The first meeting hall of its kind in San Antonio and still in use today, it was ideal for conferences, large charity events, and public gatherings.

The hall's design was sometimes compared to a flying saucer. Outer space was on everyone's mind, especially after the Soviet Union's launch of its Sputnik 1 satellite. The endeavor raised questions about the vast galaxies beyond planet earth and triggered worries about

Air conditioning began to boom in the late 1950s, and new window units created a large increase in residential customers in San Antonio.

As the decade ended, the company demolished its first power plant, located at the corner of Presa and Villita Streets. Architect O'Neil Ford designed La Villita Assembly Hall, which was built on the site and operated by CPSB.

national security and whether U.S. technology was as advanced as it should be. In 1958 President Dwight Eisenhower and Congress established the National Aeronautics and Space Administration. The simple preamble for its establishment describes "an Act to provide for research into the problems of flight within and outside the earth's atmosphere, and for other purposes." Over the next decades the country's space program would make extraordinary advances, including putting a man on the moon; "other purposes" would give its activities immense flexibility. In the early years NASA combined the electric profession with science and national security—an important union that would endure into the future. By the decade's end more than a third of the country's scientists and engineers in universities were working on government research, mainly defense projects.

Braunig announced his retirement in 1958, after nearly nine years as the company's general manager. He boasted that he knew each one of the almost 1,850 employees by first name. Now worth $132,400,000—nearly $100 million more than its purchase price—the company had grown at a dizzying pace. Two plants, a high-voltage transmission system, a rebuilt distribution system, a supervisory control system, and an office building had been added. The Villita Assembly Hall was almost complete, and a third 100,000-kilowatt generating unit was almost operational at the Mission Road plant, driven by steam from a Babcock & Wilcox boiler capable of evaporating 750,000 pounds of water an hour at a temperature of 1,005 degrees Fahrenheit. It had flue stacks over 150 feet tall and a 175-foot-tall water tower that held 40,000 gallons of water. The utility company's march into the future had been dramatic, and Braunig was confident that Assistant General Manager Otto Wahrmund Sommers would lead

it forward. He assured his employees, the trustees, and the citizens of San Antonio that "with its building and expansion plans, the CPSB expects to retain its position as one of the better and more progressive utilities in the United States."

Sommers was born in San Antonio in 1906, the year Tuttle arrived to oversee various utility companies owned by American Light & Traction. His grandparents were German, from Fredericksburg, Texas, and his father was a lawyer who dabbled in San Antonio real estate until he bought a drugstore at the corner of Alamo Plaza and Houston Street in 1912. Sommers Drugs eventually became one of the largest chains in the Southwest. Young Otto attended Texas A&M University, graduated with honors in electrical engineering in 1929, and was employed by

Otto Wahrmund Sommers became general manager of the company in 1959.

SAPSCo as a cadet engineer. In that position he was part of the substation crew, did relay work, and held positions in various departments as he climbed the corporate ladder.

The company's success would continue, but the decade would also bring significant challenges to Sommers, his team, and CPSB trustees, as the complexities of a modern power industry followed the boom of the postwar era.

In 1951 an amendment had been made to the CPSB trust indenture to secure additional bonds, taking advantage of very low interest rates and authorizing additional payments to the City's general fund. This was controversial, of course, and there were varying opinions on the board. Gen. John Bennett and Mayor Jack White were strongly in favor, arguing that payments should reflect the growth the system enjoyed. Bennett suggested a number—14 percent of the CPSB's gross earnings. The two bankers on the board—Joe Frost and Walter Napier—were against it, and Willard Simpson was uncertain. Bennett worked tirelessly to get a majority and the board finally approved the amendment. According to a 1992 memo written by Bennett, "one of the two bankers immediately rose to his feet and announced that he was resigning from the board. In spite of personal calls later, this individual would not change his vote." James H. Calvert was elected to fill the vacancy created by Frost's resignation.

In 1959 the city council urged CPSB trustees to cooperate with the 1951 amendment and authorize additional payments to the general fund. The increase required approval by the holders of 75 percent of the outstanding bonds. Once approval was secured, the indenture was executed on February 1, 1960, formally putting increased payments into effect. This change would create what has been described as "a golden goose for San Antonio," providing more than $6 billion in revenues over the next fifty-seven years to support the city's infrastructure and entrepreneurial growth. The revenues received from the municipally owned utility company are greater than the city's annual property taxes and have powered San Antonio's dreams for the future.

In 1960 San Antonio celebrated one hundred years of gas distribution. Mayor John Edwin Kuykendall presided over a great deal of public fanfare. A twenty-foot ceremonial torch was lit in front of the Alamo in honor of the pioneers who introduced gas to the city in 1860. Situated at Houston Street on the banks of the San Antonio River, the first gas company had manufactured its product from resin shipped from the Northeast to the port of Indianola and delivered to San Antonio by oxcart. In 1907 the plant moved to Salado Street, and gas was made by passing steam through incandescent hot coke or anthracite and adding oil to enrich it. It remained an expensive product, a luxury for the most part, until the discovery of natural gas in South Texas in 1922.

Almost immediately after that discovery, Southern Gas built a pipeline supplying San Antonio with natural gas. In 1930 the United Gas Public Service Company assumed control of Southern Gas, and although it eventually changed its name to United Gas Pipeline, the same company continued to supply San Antonio with its fuel through a contract that would expire in 1962. It had recently built a large belt line that encircled the city, allowing CPSB to distribute gas to its 158,774 customers from various points. As its forty-year contract approached expiration, the City asked for bids, and six pipeline companies began to work on their proposals.

The CPSB anticipated a good year as the 1960s began, but it recognized that with increases in the cost of labor and material, the price of expansion would climb. Financing construction would come largely from operating revenues, supplemented by $5 million from the revenue bonds sold in 1951. Some trustees completed their terms and new ones were appointed. As the decade began, James H. Calvert, Leroy Denman Jr., John M. Bennett Jr., Melrose Holmgreen, and newly elected Mayor Walter W. McAllister Sr. prepared to take the organization into the future.

Calvert, who had served on the board since 1951, was elected chairman. Born in Chester, England, he served in the Royal Air Force during World War I and came to Boston in 1920 to work for Allied Stores. When the chain acquired Joske's in 1932, Calvert was sent to San Antonio to revive the city's first department store. Established by German immigrant Julius Joske in 1867, under Calvert's leadership it became the

first store to be fully air-conditioned in 1936, and the first to offer integrated dining areas in 1957. Well-connected in the community, Calvert was a former chairman of the Bexar County Red Cross and former president of the San Antonio Chamber of Commerce; he served on the boards of Alamo National Bank, Trinity University, and the Southwest Research Institute.

Denman, a new board member, was an attorney and banker, chair of the board of San Antonio Loan & Trust, and a director at Frost National Bank. During World War II, the U.S. Department of State sent him to South America to investigate concerns about growing Nazi activity there; assignments in Guatemala, Washington, D.C., and Argentina followed. When Denman returned to San Antonio he joined his uncle's law firm and became involved in diverse philanthropic organizations.

The CPSB trust indenture called for independent board leadership that avoided politics.
In 1961 that leadership included (left to right) Mayor Walter McAllister, General Manager Otto Sommers,
board chair James Calvert, Gen. John Bennett, Leroy Denman Jr., and Melrose Holmgreen.

Bennett, who had been on the board for a decade, had an impressive military background as a combat bomber pilot in World War II. Stationed in England as part of the 8th Air Force, he led twenty-seven bombing missions over Germany and received a Silver Star, Legion of Merit, Distinguished Flying Cross, Bronze Star, Air Medal, and the Croix de Guerre with Palm, eventually earning the rank of major general. He had pushed hard for the change to the trust indenture the previous year, in favor of providing more money to the city, and in 1958, as chairman of the National Bank of Commerce, he oversaw the design and construction of the first skyscraper built in the city since the glory years preceding the Great Depression.

Holmgreen was president of Alamo Iron Works, which his grandfather established in 1878 on the banks of the San Antonio River. The foundry operation supplied steel for early projects like the Menger Hotel and became known for its production of manhole covers. By the 1960s it was producing industrial supply products, including paints, lubricants, and welding equipment. Brought on the board by his close friend Tuttle, Holmgreen was one of the original trustees of the San Antonio Medical Foundation, which acquired the land for the South Texas Medical Center, a large medical complex in the city's northwest quadrant.

McAllister succeeded J. Edwin Kuykendall as mayor in 1961. His grandfather had come to Texas in the 1840s and was the first county judge at the Bexar County Courthouse, designed by architect James Riely Gordon in the late 1800s. McAllister earned a degree in electrical engineering from the University of Texas and worked briefly as a surveyor for Colorado River & Power. McAllister was an entrepreneur, establishing the

San Antonio Savings Association with Henry Halff in 1921. President Eisenhower appointed him chairman of the Federal Home Loan Bank Board in 1953. After his appointment to the city council in 1960 to fill an unexpired term, he was elected mayor, serving five two-year terms, until 1971.

The trustee appointments reflected the founding tenets that the organization be led by businessmen who actively supported their community and who were not linked to politics. They were more than qualified to meet the challenges ahead. The first challenge occurred when a backlash against gas meters spread through the country, gravitating to Houston and then to San Antonio. Media and citizens protested the use of new "demand meters" that measured electricity used and recorded peak consumption, allowing utility companies to charge more for peak usage. Only about 5 percent of CPSB's customers had demand meters, but that seemed irrelevant to the public and the outcry grew. The board followed the lead of other utilities and did away with the meters, with lost revenues of $150,000 a year. Around the same time CPSB made a decision that would have a far greater impact on its future.

Six companies had submitted bids for a new gas supply contract: longtime provider United Gas, Houston Pipeline Company, M. M. Cherry & Associates, Coastal States, Intra-Tex Gas Company, and Alamo Gas Company, which was headed by Glen A. Martin with local investors including R. F. Schoolfield, George A. Coates, O. R. Mitchell, George Parker, and Fred Shield. Alamo Gas, the low bidder, was awarded the contract in June 1961. The second lowest bidder was Houston Pipeline, followed by United Gas.

The contract was to go into effect on April 1, 1962, upon the expiration of the contract with United Gas. But United Gas challenged the power of CPSB and the City to enter into the new contract, filing a petition before the Texas Railroad Commission, the state agency responsible for gas infrastructure and oversight. The petition was denied; United Gas appealed in the District Court of Austin, and the case went to trial. Judge Jack Roberts ruled in favor of United Gas. CPSB attorney Matthews took the matter to the Court of Civil Appeals, and the Railroad Commission's verdict was upheld. This took many months. While the litigation was in progress, Alamo Gas proceeded to build pipelines to the gas fields in South Texas. It also sold part of its capital stock to Oscar Wyatt Jr. and Coastal States. By April it was supplying San Antonio's gas for a little less than twenty-two cents per thousand cubic feet. Large quantities of dependable, cheap natural gas made this price possible and would fuel the city's growth throughout the 1960s.

Also fueling the growth was the biggest budget in CPSB history, totaling $22 million. The largest expense was the 165,000-kilowatt generating unit being installed at the Tuttle plant. Another big budget item was the $4 million completion of the East Lake dam in anticipation of a power plant named for Braunig, scheduled to be operational by 1965 or 1966.

Gas pipeline installation had been greatly modernized by the 1960s.

By summer veteran pipeliners were racing to complete the last twenty miles of CPSB's sixty-five-mile gas loop on the city's north side before cold weather struck. The loop line on the east and west sides had been completed in March, and the connections with the Alamo Gas pipeline had been made. The August heat made the last work especially grueling. As each section was completed, sometimes after dynamiting through rock, the twenty-four-inch pipe was delivered in forty-foot sections. Cranes hoisted it into place, and the joints were welded together and x-rayed to check for imperfections. The pipe was wrapped to provide cathodic protection and lowered into a trench, which was then backfilled. After being tested with 400 pounds of compressed air pressure, the pipe became a new gas transmission line.

Calvert's term ended in 1962, after he had served as a trustee for more than a decade, and Albert Steves III was appointed. A descendant of John Smith, the last messenger to leave the Alamo in 1836, Steves was the proprietor of Ed Steves Lumber Company, established in 1866

Construction began in 1962 on the Braunig plant, fifteen miles south of the city. Its man-made lake would use recycled water from the San Antonio River in the first plant cooling system of its kind in the country.

as the city's first lumber company and supplying a market stretching to northern Mexico. His great-great grandfather had been San Antonio's first mayor; his grandfather, Albert Steves Sr., was mayor in 1912; his maternal grandfather, Sam Bell, was mayor in 1917; and his brother, Sam Steves, was mayor in 1952. Of German and Canary Islander descent, Albert Steves was born in 1907, attended Episcopal High School in Virginia, and graduated with degrees from Washington and Lee University and Harvard University. He returned to San Antonio to work in the family business. In World War II Steves participated in campaigns in Normandy, Belgium, Holland, and Germany, earning five battle stars. After the war he became president of Steves Industries, comprised of several family businesses, and when those were split in 1954 he continued to run the lumber company. A director of Alamo National Bank, the San Antonio Symphony Society, the San Antonio Zoological Society, the Southwest Foundation, and the San Antonio Medical Foundation, he—like everyone associated with CPSB—was in step with Tuttle's

The first water for the thirteen-acre Calaveras Lake was pumped from the San Antonio River in 1962.

vision of the utility as a partner in the city's development.

The community would need to be reminded of this partnership and CPSB's deep commitment to customers when rumors of the first proposed rate increase since 1936 began to circulate. Bennett, now chairman of the board, noted that "even with higher rates, San Antonio will continue to enjoy both gas and electric rates as low as or lower than most major cities in the United States. They will be lower than any of the present or proposed rates in any of the cities in Texas."

As the year drew to a close, at a remote spot on the San Antonio River about fifteen miles south of the city, the 1,350-acre East Lake received its first water, shot into the reservoir from the river. Engineers estimated that they would pump excess river water for about two years to fill what would become the county's largest body of water and the cooling lake for the Braunig plant being designed near Elmendorf, Texas. Cooling lakes, an innovative idea developed in response to the drought of the 1950s, used river water to cool power plants in a recycled system. They were the first of their kind in the United States, pioneering the environmental movement's goal of preserving valuable water from aquifers.

Sommers assured citizens of San Antonio that the added capacity of the future plant, scheduled to come on line in 1966, and the Tuttle plant's almost completed generator, would "meet San Antonio's ever-increasing demand for electricity."

A cold front in December tested the new gas lines, and to everyone's relief no power loss

occurred, indicating that the lines would provide a dependable source of power for years to come.

The *Broadcaster* magazine celebrated forty years of publication in 1962, driving home the message that above all else, CPSB had developed a corporate culture focused on employees and customers. There were new ways for employees to become involved in the community, from the United Fund to in-house service clubs. Extracurricular activities deepened bonds between employees and management. A lineman's

Gen. John Bennett, who served on the CPSB board in the 1950s, was chairman when the first rate increase in twenty-five years was being considered.

Outdoor lighting, powered by electricity, began to trend in the early 1960s.

basketball game, staged on utility poles high above the ground, offered a dramatic glimpse of the skill and physical strength required to install power lines, and citizens and employees who watched were enthralled. The December *Broadcaster* reported that Jan Carpenter, a nineteen-year-old steno clerk in the personnel department, had been crowned Miss Kilowatt 1962 and that CPSB had installed lighting for Travis Park as a year-end gift to the city.

As customers adjusted to slightly higher utility bills, more storefronts and residences were lit with holiday panoramas. Reassured that Bennett was right about higher rates in other cities, they felt confident that San Antonio's efficient service at low rates could be expected to continue. No one imagined that a decade later dependable, cheap natural gas would come to an end.

PART 6

Power Struggles
San Antonio, 1963–1981

Utility rates and power supply were not on Sid and Lila Cockrell's minds when they got an opportunity to move to San Antonio. When Sid was offered the position of Bexar County Medical Society director, his wife contacted her uncle in San Antonio to ask about the city's quality of life. Attorney C. Stanley Banks invited his niece, her husband, and their two little girls to visit. "We came for a look-see, and I fell in love with the river and the city. And the rest is history," former San Antonio mayor Lila Cockrell recalls. The Cockrells made the move in 1955, and by 1962 the homemaker was deeply involved in the community, recognizing that San Antonio truly was a "city on the rise, largely due to the plentiful supply of power at reasonable rates."

Cockrell's interest in city growth was not new. In Dallas she had served as president of the local League of Women Voters. Her civic and political activities in San Antonio soon caught the eye of Mayor McAllister. He called on her one day to see if she would consider running for city council as the first woman the Good Government League

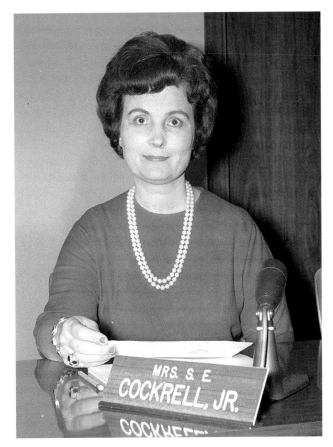

Lila Cockrell was convinced to run for city council by Mayor McAllister in 1963.

had ever supported. The city's prominent businessmen had established the organization in the early 1950s when San Antonio abandoned its governance structure of city commissioners in favor of a city council and manager. Working to shape the city's growth and political future, the league became a political powerhouse, and the candidates it promoted were almost always elected. Cockrell agreed to run and was elected to the city council in 1963.

That year Coastal States acquired all of the Alamo Gas stock, just a year after the San Antonio–based company had been awarded a twenty-year contract with CPSB to supply the city with natural gas. It established a subsidiary, LoVaca Gas Gathering Company, and began increasing its control of gas pipeline facilities in San Antonio, Austin, and South Texas. Over the next decade it would acquire 965 miles of gathering and transmission lines from United Gas, extending from Laredo to San Antonio to Austin and providing service to more than fifty Texas towns. It would also acquire lines from Texas Gas & Utilities, Nueces Industrial Company, and other smaller companies that served towns in South Texas and the Corpus Christi area.

San Antonio was a busy place on many fronts, and the impact of Coastal States' expansion was not much noticed. But there were an increasing number of media stories about the meteoric rise of Oscar Wyatt Jr., its colorful and dynamic CEO. Wyatt was born in Beaumont, Texas, in 1924. His father, an alcoholic, abandoned the family, and Wyatt and his mother moved to the little town of Navasota where he worked as a crop duster to earn extra money during his teenage years. An excellent student, he was accepted at Texas A&M University but left after a year to serve in World War II. A combat bomber pilot,

Oilman Oscar Wyatt and his wife, Lynn, were favorite characters for an increasingly curious media audience.

Wyatt was wounded twice during battle and was a decorated aviator before the age of twenty-one. After the war he returned to Texas A&M and earned a degree in mechanical engineering. To make money during college, he sold drill bits from the trunk of his Ford Coupe to small oil companies and later worked for Kerr-McGee and Reed Roller Bit. He became a partner in Wymore Oil in 1954. A year later Wyatt founded Coastal States with just sixty-eight miles of pipeline and seventy-eight employees. The company expanded rapidly through acquisitions and began diversifying its holdings, and by 1960 its revenues exceeded $17.6

million. When Wyatt married Houston socialite Lynn Sakowitz in 1963, the self-made mogul's visibility increased and he became even more interesting to the media and a public that delighted in larger-than-life characters.

Cockrell, who was observant, was slightly worried about Wyatt. She asked Mayor McAllister if she could attend the CPSB board meetings, curious to know more about her city's power supply and future reserves. She was told that board meetings were closed, a policy she changed when she became mayor a little over a decade later. She continued to watch the city's partnership with Alamo Gas / Coastal States / LoVaca with a careful eye.

In the early 1960s a group of enterprising business and civic leaders organized by banker Bill Sinkin began to promote the idea of bringing a world's fair to San Antonio in celebration of the city's 250th birthday and to attract tourists and businesses. H. B. "Pat" Zachry, Marshall Steves, Tom Frost, Jim Gaines, Jessica Hobby Catto, Flora Cameron Kampmann, Dr. Joaquin Gonzalez, John Steen, Nancy Brown Negley, Mayor McAllister, and a long list of citizens envisioned an international exposition that would focus on the Americas. The group worked for nearly a decade to solicit official approvals, raise money, attract exhibitors, and commission public artwork for the fair. A first step was the purchase of ninety-seven acres downtown, where construction would begin a few years later. Of course, everything being built would need power.

In November 1963 President John F. Kennedy was assassinated in Dallas, one day after visiting

Businessman Bill Sinkin (left) and Mayor Walter McAllister began to promote the idea of an international exposition in San Antonio in the early 1960s.

San Antonio. Martin Luther King made his "I Have a Dream" speech the next month. Television brought the Vietnam War into America's living rooms with unprecedented dramatic reality. Despite the city's strong economy and sense of well-being, the world beyond revealed an angst and a glimpse of complex challenges ahead.

Planning for challenges in the power industry was something CPSB did well. Based on the year's utility usage statistics, Sommers predicted that San Antonio would have a population of 1 million by 1970. In preparation, design plans for the Braunig plant had been finalized, water continued to flow into the man-made East Lake that would eventually provide cooling for the plant, and a

President John F. Kennedy was assassinated in Dallas on November 22, 1963, one day after visiting San Antonio. His motorcade had traveled down Broadway, where schoolchildren lined the street to see President and Mrs. Kennedy and Gov. and Mrs. John Connally.

railroad spur was built for transport of the 297-ton generators and other equipment to the plant. Raymond "Kitty" Butler took the lead on the construction project, which would enable Union Pacific cars to access the Braunig plant from the Southern Pacific's main line. Butler described the job as "horribly challenging . . . especially the hordes of rattlesnakes and large bed of quicksand" that confronted the crews. They prevailed, and a special flat car delivered the first giant Westinghouse generator to the site. Former general manager Braunig, now general manager of the San Antonio River Authority, wielded his shovel to move the first dirt at the groundbreaking a few months later. When Houston-based construction company Brown & Root began work on the first

phase of the $17 million plant, no one imagined that the railroad spur would prove essential for transporting more than heavy equipment in the years to come. A fourth generating unit—the largest ever at 170,000 kilowatts—was installed at the Tuttle plant, enabling the utility company to meet all energy demands, including an increase in air-conditioning and the load required for interior lighting of Natural Bridge Caverns, a recently discovered cave in Comal County that was expected to become a major tourist attraction.

Freeways were extending at a rapid pace, and it was becoming easier than ever to travel beyond the city—northeast to places like Natural Bridge Caverns and northwest to suburbs where

As San Antonio's expressways expanded, CPSB installed lighting to make nighttime driving safer.

a medical center was under construction. Modern lighting made the freeways far safer after dark, as gaslights had done on Alamo Plaza.

Customers were delighted to see a reduction in their rates, thanks to less expensive electrical equipment and an important legal settlement that awarded CPSB $2.65 million. Longtime attorney Matthews and two young associates, James Baskin and Richard Goldsmith, led the CPSB suit against equipment suppliers Westinghouse, General Electric, and Allis-Chalmers for antitrust violations and bid rigging. The company passed along the spoils to its customers by reducing rates.

McAllister was reelected mayor for the third of five terms. He was justifiably proud of the report he made about the city's growth. Since the City purchased the CPSB in 1942, the population had more than doubled; the demand for natural gas had increased by 500 percent, and demand for electricity had increased by 800 percent. A service center was under construction on the city's east side, designed by architect Bartlett Cocke with H.B. Zachry as the site contractor.

Power plants and a growing number of substations were changing the city's landscape. Two substations—in southeast and southwest San Antonio—had been added to the network of what General Foreman E. F. "Charley" Townsend called "big hummers." Townsend, who had overseen their construction, described them as a "mesh of steel beams, girdles, transmission wire and cable"

By the mid-1960s the utility company's power plants had all been expanded.
The fifty-year-old Mission Road plant still towered over the San Antonio River on the southeast side.

that could be found all over the city at three-mile intervals, ensuring the reliable delivery of electricity to industries and homes.

H.B. Zachry's completion of the remaining nine miles of the horseshoe-shaped gas loop around San Antonio guaranteed security for the city's gas supply. Sommers explained to customers that, "should an interruption occur in Alamo Gas's lines, damage can quickly be isolated by shut-off valves and the remainder of the system can be served from another source."

Two odorizing stations had been added to the system, infusing natural gas from South Texas with a chemical that gave it a slightly sweet scent and making gas easy to detect. CPS had odorized its gas since 1937, when the nation realized that undetected gas leakage could result in tragedy. More than two hundred children died in a school explosion in New London, Texas, when leaking gas went undetected because it had no smell. State laws, and eventually federal laws, were changed requiring gas to contain a warning agent. Over the years Calodorant B-1, manufactured by Oronite Chemical Company in Shreveport, Louisiana, had become the agent of choice. It was shipped by truck to San Antonio, and CPSB added it to the gas it purchased from LoVaca.

In addition to concern for its customers, CPSB had made employee safety a top priority from its earliest days. SAPSCo and the small utility companies that preceded it had always emphasized accident prevention, offering safety training, a reward system for following safety procedures, and ongoing research about improvements to equipment and clothing. In 1965 the company tested "aerial baskets" for its linemen, who now wore hard hats and gear that would have amazed the men who brought electricity to the city more than six decades before.

That year Phelps Construction completed a garage for CPSB's transportation department, which owned 800 vehicles of varying types. The Salado Street Garage provided maintenance and repairs to the growing fleet and was designed to handle expansion as plans for the company became realities.

The Victor H. Braunig plant's opening in 1966 signified a major addition to the company's growing electric system. In addition to providing power, it offered recreation opportunities; its man-made lake was stocked with red drum, catfish, and largemouth bass and had picnic and camping sites.

By the following year, with much of the funding for HemisFair in hand, San Antonio was experiencing another building boom in preparation for an influx of tourists. H.B. Zachry built the Hilton Palacio del Rio on the city's River Walk, which had undergone a massive upgrade including sparkling lights and a row of restaurants. The hotel was built in a record 202 days using lift slab construction, developed at the Southwest

Over the years employee safety became a top priority and equipment underwent massive improvements, including state-of-the-art hard hats and high-technology buckets for linemen.

Research Institute, in which floors were lifted into place and entire rooms were positioned inside. A convention center was under construction, with a 25,000-square-foot mural commissioned to cover one side that faced the river. The mural, Mexican artist Juan O'Gorman's *A Confluence of Civilizations*, was designed to capture the theme of the fair and San Antonio's emerging identity as a place where cultures coexisted. In the same spirit of harmony, CPSB arranged for the roof of the Villita Assembly Hall to be painted like a flower

The Braunig plant opened in 1966, providing customers with both electricity and opportunities for recreation around its lake.

because it would be visible to HemisFair tram riders. The skyline changed even more dramatically when architects O'Neil Ford and Boone Powell designed the Tower of the Americas as the fair's landmark structure.

First Lady Mrs. Lyndon Johnson, U.S. Congressman Henry B. Gonzalez, and Mayor McAllister opened the fair on April 6, 1968, and over the next six months more than 6 million visitors toured the ninety-six acres of pavilions representing more than thirty countries. General Electric, Eastman Kodak, Southwestern Bell (later AT&T), IBM, Humble Oil (later ExxonMobil), American Express, and other corporations sponsored exhibits, set against a backdrop of

fountains and public art commissioned by local philanthropists. Performers like Los Voladores de Papantla—the "flying Indians from Mexico"—and international food and music contributed to a celebratory spirit.

Another celebratory moment occurred in June 1969, when Gov. Preston Smith signed House Bill 42 into law, establishing the University of Texas at San Antonio. State Sen. Frank Lombardino and civic leader John Steen were instrumental in getting the legislation passed. In the years that followed, UTSA would strive to become a Tier One research institution and the largest university in the San Antonio metropolitan region.

The celebratory spirit was dampened when San Antonio experienced a few minor interruptions in gas service that winter. CPSB asked Coastal States to provide an accounting of their supply contracts and reserves, wanting to be sure that the "ample reserves" guaranteed by the gas supplier existed. Coastal States refused, claiming it was private information. With construction under way on two large diameter pipelines from gas fields that had been discovered in Reeves, Ward, and Pecos Counties, the company controlled pipelines that created a monopoly for the gas supply to a major part of Texas.

As always, CPSB followed a preparedness strategy, exploring a fuel diversification program so that it could draw on alternative energy sources should problems arise with the natural gas supply. The company joined power companies in Houston and Austin to conduct a feasibility study for a nuclear power plant and began to look again at coal as a fuel source. It planned to expand the

When HemisFair opened in 1968, VIPs including (left to right) U.S. Congressman Henry B. Gonzalez, First Lady Mrs. Lyndon Johnson, and Mayor Walter McAllister attended the festive ceremonies.

Gov. Preston Smith (center) signed the bill to establish the University of Texas at San Antonio in front of the Alamo in 1969. State Sen. Frank Lombardino (left) and civic leader John Steen (right) were strong proponents.

Braunig plant, and a new plant—the state's first totally gas-fired plant, named in honor of retiring General Manager O. W. Sommers—would open in 1972. Still another, even larger plant was being designed on an 8,000-acre site southeast of the city. Its dam had already been built, creating Calaveras Lake, which would cool the plant with recycled wastewater when it became operational—one of the first projects of its kind in the country. Like Braunig Lake, Calaveras Lake would offer recreation, eventually becoming a major community site for CPS's youth programs.

Anticipating that electricity usage in San Antonio would increase by 11 percent over the next twelve months, the utility analyzed population and usage data to prepare for the next decade and built big reserves into its designs. The newest plants would generate more than 5,700 megawatts of electricity, maintaining a 15 percent reserve as far out as 1985.

J. Thomas Deely, who succeeded Sommers as general manager in 1971, had worked for the company for thirty-five years. His special sensitivities to rates from both the provider and customer perspectives made him the perfect person for the job. Adversities in childhood had shaped a strong, resilient leader, and these traits would be useful during the looming energy wars. His father, a master tailor in Comanche, Texas, had been killed by an explosion in his shop in 1912, when Deely was two years old. His widowed mother raised him, instilling a can-do spirit and a penchant for hard work. Deely mowed lawns, jerked sodas, cut brush, picked cotton, and threshed grain to help pay the family bills. In 1928 he entered the University of Texas but was forced to drop out in 1929 because of the Great Depression. He accepted a job with the Brown

J. Thomas Deely became general manager of CPSB in 1971.

County Water Improvement District surveying and mapping the building site for the Brownwood Lake dam. He returned to the University of Texas in 1932 and graduated with honors the following year. The Texas Highway Department hired him as a transit man and promoted him to office engineer, where he developed road plans. At night and on weekends he worked for the Works Progress Administration making maps and surveying work projects. In 1936 SAPSCo was recruiting engineers, and the University of Texas recommended their graduate Deely. He started as an industrial engineer and was eventually put in charge of the rate division, excelling in statistics and moving the company into the world of computer technology.

Over the years Deely had worked closely with Braunig and Sommers, who mentored him on planning for the city's future energy needs.

A visit from Oscar Wyatt to San Antonio in early 1971 raised concerns about whether Coastal States had been as careful in its planning. In a conference with Mayor McAllister and three other CPSB board members, the oilman maintained that Coastal States had the reserves to supply San Antonio's gas for the rest of the contract but encouraged CPSB to negotiate a new contract at a higher rate—to ensure a gas supply for the years after the contract ended. The meeting raised apprehensions, and CPSB consulted with its attorney. Matthews advised his clients that they were entitled to information from Coastal States, and they made a written demand for data about the gas reserves dedicated to San Antonio.

Coastal States and its subsidiary LoVaca immediately filed suit in the state district court of San Antonio, seeking a declaratory judgment that the City and CPSB "had no right to audit or verify Coastal's gas reserves or its commitments to customers against such reserves." Wyatt began to ask for conferences to work on a settlement, and CPSB appointed its newest board member, John Newman, whose business was oil and gas, to pursue a solution with Deely and Matthews. Coastal States' senior vice president, thirty-seven-year-old William E. Greehey Jr., was determined to find a solution as well. The group met more than twenty times over the next few months. Wyatt continued to urge CPSB to renegotiate its contract at a higher rate for gas, insisting the rate was necessary to provide the city with gas after 1982, when the current contract would expire. Newman understood Coastal States' economic problem of buying new gas at a higher price but maintained that CPSB needed full knowledge of

the company's low-cost reserves before it could consider another contract. Wyatt continued to press for cancellation of the existing contract and for a long-term, cost-plus contract. About this time other cities receiving gas from Coastal States, including Austin and Corpus Christi, joined in the push for information about its gas reserves. As media coverage of the energy crisis expanded, reporters and others began to refer to the utility company as CPS (City Public Service) instead of CPSB (City Public Service Board), perhaps to differentiate between the company and its board of trustees, which found itself in the spotlight too. The shorter name stuck.

In 1971 John Gatti was elected mayor, succeeding McAllister. Gatti was at the city's helm when CPS celebrated thirty years of municipal ownership. Media accounts extolled the growth of both the company and San Antonio. Its value had increased more than 1,000 percent, to $350 million, by 1972. CPS revenues had poured into San

When John Gatti was elected mayor in 1971, he became a member of the CPSB board.

Antonio's infrastructure, keeping citizens' property taxes lower than those of most major cities.

Gatti encouraged Tom Berg to join the CPS board, assuring him that he would enjoy the experience and "meet some mighty fine people. There isn't any trouble at CPS; its board meetings are not time-consuming, and pretty much routine." Berg, who had moved to San Antonio from New York in 1967, was president of Friedrich Refrigeration and had recently overseen the construction of a 500,000-square-foot plant. He agreed to fill the board vacancy. In his oral history for the University of Texas Institute of Texan Cultures, he laughingly recalled that "it was all very easy, until I was elected chairman a few years later. Then suddenly there was a gas shortage. Immediately!" The oil crisis of 1973 was inevitable, a direct consequence of the country's peak oil production in late 1970 and the oil embargo imposed by the Organization of Arab Petroleum Exporting Countries, in response to America's decision to resupply the Israeli military during the 1973 Yom Kippur War.

In the first thirty years of CPS's operation, the cost of natural gas had varied slightly, between 16 cents and 23 cents per thousand cubic feet. With Coastal States' unwillingness to share information, Deely and CPS trustees were growing doubtful that the gas provider could fulfill the remaining years of its contract at the fixed price. City leaders understood that available power at favorable rates was essential to attracting industrial and residential growth and that a rate increase was inevitable in the not too distant future.

When the $35 million Sommers plant opened in 1972, the rising cost of natural gas dampened some of the fanfare that would have accompanied its dedication. Nonetheless the addition of the

eleven-story structure brought total generating capacity to 2,144,000 kilowatts—about a million more kilowatts than fifty years earlier.

Deely was determined to find ways to increase efficiency. Later that year CPS unveiled the Gas and Electric Operations Center—the first of its kind in the nation—utilizing massive computers to help gas and electric systems operate more efficiently by responding to supply and demand. Deely was hopeful that this step would bring increased reliability, and eventually savings, to customers.

Meanwhile talks with Coastal States continued. City councilman Charles Becker, a friend of Wyatt's, urged the City to negotiate the new cost-plus contract. Councilwoman Cockrell did not agree. "From my debating days, I knew that for everything that sounds wonderful, there is always another side to be found. That tempered my leaping to an opinion before all the facts were known. I always asked, what is in the public's interest? And I wanted more information."

Gatti served one term as mayor. Becker succeeded him in 1973 just as debate over the Coastal States contract began to spill out of the city council and the CPS boardroom. In December and again in early 1974, during a winter that was not severely cold, Coastal States defaulted on its gas deliveries, meeting little more than half the demands of its 400 pipeline companies. Gas was rationed under order of the Texas Railroad Commission. High gas prices impacted the costs of electricity generation and basic gas service, despite the contract with Coastal States / LoVaca. Citizens of San Antonio were not happy.

"Utility rates went sky-high overnight," Berg recalled. "When you used to pay fifty dollars a

The Gas and Electric Operations Center, the first of its kind in the country, utilized massive computers to help CPS systems operate more efficiently.

month, suddenly you were paying double or triple that for electricity. Some people were paying as much for their utilities as they were paying for rent. It was a terrible time, and the public and newspapers blamed CPS, even though we had no control over it. The 1973 Arab-Israeli War created a huge oil shortage, because of the [Organization of Arab Petroleum Exporting Countries] embargo, and most of Texas's gas was sold to states in the Northeast, at higher prices, shipped via that giant pipeline." Stories about the crisis dominated the newspapers, and public outcry grew louder.

"I said, 'We have got to sue!' Mayor Becker was reluctant. He thought Oscar Wyatt was a very smart man and that our utilities were in good hands," Cockrell recalled. "But I kept pushing, and

we finally sued in May 1974. Eventually we won, and San Antonio got much more than the rebate we passed along to our citizens."

Berg had hoped CPS could negotiate a compromise with LoVaca, but after many frustrating meetings with the Railroad Commission, which at first maintained LoVaca's right to charge cost-plus for the gas it supplied, he agreed with Cockrell that a lawsuit was inevitable. Newman left the board that year, and banker Glenn Biggs was appointed to fill the position. Robert V. West II, CEO of Tesoro, was appointed to replace long-time trustee John R. Locke, a prominent attorney who rotated off the board after ten years of service. West recommended that CPS hire William E. Miller, a sixty-seven-year-old senior partner at the Washington, D.C.–based law firm Steptoe

When the energy crisis hit San Antonio in the early 1970s, Oscar Wyatt and his Coastal States company found themselves in the middle of the fray.

& Johnson, who had successfully represented the Slick Oil Company over the years. The legal battle gained intensity.

Although the suit's outcome was eventually positive for San Antonio, the bitter fight was intense and stressful. Many friendships between smart, dedicated public servants and business leaders ended in anger, and the decade's energy wars were keenly felt.

While the suit was ongoing, the City of San Antonio decided to join Austin and Houston in a decision to move forward with the South Texas Project, the first nuclear power plant in the state. Brown & Root was selected as the architect and construction company. Diversifying still further, CPS built fuel oil tanks at its power plants to hold gas purchased directly from producers, and two

coal-fired generating units were designed for the site next to the Deely plant where construction was under way. The railroad spur that had brought equipment to the Braunig plant could deliver coal provided by the Cordero Mine in Wyoming.

San Antonio was not the only place experiencing power struggles brought about by escalating gas prices. On May 16, 1974, the New York legislature voted to buy two partially completed power plants from Consolidated Edison for $500 million each. The move rescued the utility—the largest privately held firm in the nation's biggest industry—from its immediate financial woes. According to Mark Green, writing about the purchase a few weeks later for the *New York Times*, Consolidated Edison's board chair, Charles F. Luce, hinted that his firm might have to sell its

Forecast
Clear to partly cloudy and hot Tuesday. The high Tuesday in the upper 90s after an overnight low in the low 70s. Temperatures Monday ranged between 97 and 70. Details, Page 9A.

MONDAY'S TEMPERATURES

5 a.m. 74	10 a.m. 87	3 p.m. 97	8 p.m. 92
6 a.m. 75	11 a.m. 91	4 p.m. 96	9 p.m. 89
7 a.m. 72	Noon 92	5 p.m. 95	10 p.m. 87
8 a.m. 79	1 p.m. 94	6 p.m. 95	11 p.m. 84
9 a.m. 83	2 p.m. 96	7 p.m. 94	Midnight 82

San Antonio Express

SPORTS FINAL
F

Tuesday, July 23, 1974   One of America's Great Morning Newspapers   **10¢**

Spurs Sign Freeman
See Story, Page 1D

$150-Million Suit Is Filed

Coastal Hit For Damages

By DEBORAH WESER

City Public Service late Monday afternoon filed a damage suit for more than $150 million against Coastal States Gas Corp.

The move followed an emotionally charged public trustees meeting and appears to be a major victory for CPS Chairman Tom Berg.

The vote to proceed with the suit was 4-0. Trustee Glenn Biggs was absent.

Berg has consistently pressed for the suit despite Mayor Charles Becker's insistence that court action is "premature."

Becker Votes

Although Becker said again Monday he feels the suit is premature, he called for the vote to proceed with the suit.

His move came following a five-minute recess after a long, heated discussion of the suit and Becker's contention CPS attorney Wilbur Matthews has a "conflict of interest."

Becker, crediting City Councilman Cliff Morton with suggesting a "compromise," called for the vote with the proviso that City Atty. Crawford Reeder would handle the suit.

Morton apparently made the suggestion during a brief over-the-shoulder conference with Becker during the meeting.

Attorneys from the firm headed by Wilbur Matthews will be used by Reeder as consultants, Berg noted. They will be paid for their services at normal rates.

In a surprise move only an hour after the CPS meeting, Reeder...

CITY PUBLIC SERVICE BOARD CHAIRMAN TOM BERG
. . . says damage suit against Coastal and Lo-Vaca should be filed

After intense debates among city leaders, CPS trustees, customers, and the media, San Antonio filed a lawsuit against Coastal States in 1974.

entire system to New York State, giving rise to speculation that other private electric companies squeezed by escalating fuel and capital costs might also be in trouble. Green noted that Larry Hobart of the American Public Power Association had asked rhetorically, "Con Ed as a local municipal? It could give us a black eye."

For the first time in more than a decade, CPS asked the city council for a rate increase. Eloy Centeno, a board member since 1968, appeared before the council in July to make the request, noting that the company had tried various strategies to avoid raising rates. Centeno, who owned five supermarkets, a meatpacking plant, a theater, and other properties, had worked to unify the city's Anglo and Mexican American communities during some difficult years in the 1960s and 1970s. The presence of Centeno, CPS's first Latino trustee, made a statement that the utility intended to reflect the city's demographics.

The following year San Antonio became the first of the nation's "top ten cities" to elect a woman mayor. Lila Cockrell took the helm of city government at a time when utilities nationally were being challenged by rising fuel costs and overloads. "I remember being at a banquet for the Texas Municipal League, attended by mayors from around the state, in San Antonio's convention center. The highlight was to be a performance by Rosemary Clooney, but when she stepped up to the microphone to sing, the lights flickered then dimmed to not quite dark, and the sound system went out. It all came back up after a few minutes, but the next morning the mayor of Austin brought a kerosene lantern to my office as a joke," Cockrell said. "We knew then that we needed to keep planning ahead for our energy needs. CPS has always done that."

Following up on one of her major campaign promises, Cockrell pressed the city council to make the utility company's board meetings open to the public. She recognized that openness was especially critical now that an energy crisis was on everyone's minds. At first the trustees, chaired by Berg, resisted, and the media publicized the ongoing debate, but by fall the thirty-three-year tradition of closed meetings was ended. CPS's first open board meeting drew hundreds of citizens to the company's headquarters.

Berg and Cockrell became strong allies in the push to collect damages from Coastal States and provide dependable energy to the city at reasonable rates. Newspaperman Jim Berg describes the relationship that developed between his father and Cockrell as "wonderful and full of trust." He explained that Cockrell authorized Berg to directly negotiate an exit from the multicity lawsuit against Coastal States that included Austin,

During the energy crisis, Eloy Centeno, a CPS board member since 1968, had the difficult task of asking the city council for the first rate increase in more than a decade.

Corpus Christi, and Houston, with no end of litigation in sight. Once San Antonio was operating alone it began to move closer to an extraordinary settlement that, in Jim Berg's words, "allowed CPS to produce the biggest golden egg ever . . . it was as big a story as the birth of CPSB in the 1940s."

As the year ended, the Nuclear Regulatory Commission granted a construction permit for the South Texas Project, located between Bay City and Palacios, and owned in partnership by CPS, the City of Austin, Houston Lighting & Power,

Council delivers on campaign pledge

BEGINNING OF THE END FOR CPS AUTONOMY

CAMPAIGN PROMISES: City Council this week set the stage for control of City Public Service in the future . . . it won't happen before 1984 when old series bonds fade away and the rules that govern them, too . . . but it's written into the new indenture council authorized last week . . . last minute balking by some council members, notably Mayor **LILA COCKRELL** and Councilman **PHIL PYNDUS** both GGLers, prompting Independent Teamer **BOB BILLA** into a soliloquy: "I ran for council on the platform that this council would assume control. It's a commitment I made. I would like to stand by it." . . . so much for all you campaign promise doubters.

IT REALLY HAPPENED, TOO: "I didn't realize we could," admitting Mayor **COCKRELL** and other council members . . . their consultant, local bond attorney **RICHARD HARRIS** gently showing them the law says they can design CPS and its board, or no board, any way they want . . . it's all theirs to work with . . .

LILA, opting for retaining a board—

Councilmen didn't even realize how far they could go

TOM BERG: Don't do it.

LILA COCKRELL: Sorry, Tom.

When Lila Cockrell was elected mayor in 1975, she opened CPS board meetings to the public.

and Central Power & Light. The San Antonio city council voted 7 to 1 for the new plant, and CPS trustees were unanimous in their approval. In a message to customers, Deely announced, "We are building this plant because it will produce electricity at one-fourth the cost of using natural gas or oil, and at one-half the cost of coal." He also reminded customers that two coal units were being built for the Calaveras plant, large enough to supply one-half of the city's electricity needs.

When Deely retired in 1975, the CPS board voted to name the plant by Calaveras Lake in his honor, recognizing that he had faced some of the company's most difficult years. Jack T. Spruce became the general manager. A graduate of Texas A&M University with a mechanical engineering degree, Spruce had worked for the Army Corps of Engineers during World War II and had been part of CPS for nearly thirty years. Like the leaders who preceded him, he believed cooperation was an essential part of the company culture and outlined three challenges he was prepared to face: the ongoing problem of long-range fuel supply, the company's relationship with the community after recent rate hikes and questions about San Antonio's plans to use nuclear energy, and tensions brewing between the CPS board and the city council. He vowed to tackle the increasing unrest about who should exercise control over which part of utility operations.

As public discussions about nuclear energy continued, CPS brought legendary nuclear physicist Edward Teller to San Antonio in 1976 to chair its symposium, "Energy: Key to the

Future." The Hungarian scientist had emigrated to the United States in the 1930s and was an early member of the Manhattan Project, charged with developing the first atomic bomb. In the 1940s he worked on fusion-based weapons with J. Robert Oppenheimer at Los Alamos Laboratory, resulting in the development of the hydrogen bomb. Teller had been in favor of demonstrating the weapon rather than using it, hoping that would send a message strong enough to end World War II. But a secret committee that included Oppenheimer thought otherwise, and President Truman ordered the bombs to be dropped on Japan in August 1945. Teller's relationship with Oppenheimer deteriorated, and in 1954 he testified against his former mentor at security clearance hearings in Washington, D.C. He cofounded the Lawrence Livermore National Laboratory in 1958 and became a tireless advocate for a nuclear energy development program that would include a strong nuclear arsenal and nuclear testing. When he came to San Antonio for the conference, he had just stepped down at Lawrence Livermore to lobby against the proposed nuclear test ban that seemed to be gaining public support. With the rise of environmentalism in the 1970s, the antinuclear movement was growing, and groups like the Sierra Club, Friends of the Earth, and the Union of Concerned Scientists were raising questions about the safety of nuclear power. The *Broadcaster* reminded readers that "radioactive materials have been used for more than eighty years, have greatly benefited mankind in medicine (X-rays), research, preservation of foods, and other ways." But the debate was far from over.

In late 1976 excavation in South Texas began on the first of two 1,250,000-kilowatt units, and by the end of the year steel forms were placed so that construction could continue.

Jack T. Spruce (left) became the general manager of CPS in 1975, following J. Thomas Deely's retirement.

Meanwhile CPS made staff reductions to offset its rising costs. The company did not lay off employees; they simply did not replace those who left or retired. Spruce lamented that "we've squeezed personnel down just about as tight as we can," noting that the company was 355 employees short of its authorized strength, down 100 employees from the year before. He oversaw the organization of a speakers bureau to help regain public confidence, explaining that "we have a good story and I think we should be positive about it. We all must work together to present that story with a unified front to the community."

In late 1976 Brown & Root began excavation for the controversial South Texas Project, the state's first nuclear plant, located between Bay City and Palacios.

Public confidence was not bolstered when the CPS board went to the city council in May 1977 for another rate increase. Around the same time the multicity lawsuit against Coastal States was nearing trial. Coastal States executive Greehey was

as anxious as anyone to settle the case, recognizing that the company's future could be at stake. He called a man he considered to be a brilliant problem solver. "I went to see Mr. H. B. 'Pat' Zachry," he recalled. "He gave up his entire afternoon to discuss the situation with me and outlined his ideas on a big flipchart. The first thing he wrote was 'Coastal must shed blood.' He also advised that Coastal would need to divest itself of LoVaca and came up with some other ideas for a settlement that could produce a solution."

Based on Zachry's advice, Greehey approached Matthews and his legal team, which included protégés Jon Wood and Roger Wilson. A settlement was reached, ending the more than ten-year power struggle. Coastal States agreed to pay $150 million for its breach of contract and to divest itself of LoVaca and move it to San Antonio. All of the preferred stock in LoVaca, valued at $115 million, would go to the City of San Antonio, along with 13 percent of the common stock, which represented Wyatt's interest in the company. A note for $8 million from the newly formed company was also part of the settlement trust. Coastal States also agreed to release Greehey to run the company, which was renamed the Valero Energy Corporation and eventually became a Fortune 500 international manufacturer and marketer of transportation fuels with assets of more than $44 billion, and one of San Antonio's most generous corporate citizens. CPS could finally reduce its rates, and customer satisfaction began to climb.

William Greehey, senior vice president of Coastal States, played a major role in the city's complicated energy negotiations in the 1970s.

A fixed price was also negotiated for the cleaner, extremely efficient coal mined in Wyoming that would fuel the two generating units at the nearly completed Deely plant, which towered twenty-four stories when it became operational in 1977. CPS estimated that the plant's generating units would consume 8.5 tons of coal a minute. Arrangements had been made for coal to arrive by rail every other day. Together the units would be capable of producing 60 percent of San Antonio's electricity.

To further increase energy efficiency, CPS developed a computer-operated system to assess electrical demand. The DLC system (digital load control) was the first in the nation to be developed in-house by an electric utility company. Once demand was assessed, the system decided which CPS generating unit could satisfy the demand most economically. It also ensured that CPS could maintain enough generation at all times to meet the requirements of the South Texas Interconnected System.

CPS was exploring every possible means to survive the energy crisis, and in 1977 it composed an open letter to President Jimmy Carter, who had told the nation in a recent television broadcast that the country would include "conservation, coal, and nuclear energy in its future energy plans." The letter, printed in the August *Broadcaster*, began with the words "The energy joyride is over" and concluded, "When it became obvious that our gas supplier, LoVaca, had not provided the reserves to fulfill the contractual requirements of its customers, we took other measures to insure full service."

Those measures were paying off. When the Deely plant's first unit became operational in 1977, Cockrell and Berg participated in a unique ribbon ceremony at the dedication. They pulled

A coal-fired plant named for former general manager J. Thomas Deely was dedicated in 1976 by Mayor Cockrell and CPS board chair Tom Berg.

With the focus on coal as a fuel source, a railroad car maintenance shop was built in 1978 to service the 800 utility-owned railroad cars that transported coal from Wyoming.

switches that ignited a trough of coal, and the flame burned through a ribbon and powered lights that spelled *C-O-A-L*. At the dedication Deely praised CPS's foresight in converting the plant from gas to coal during the planning stages and predicted that by 1982 coal plus nuclear power from the South Texas Project would generate 75 percent of the city's electricity. Cockrell agreed, adding that "we in San Antonio stand today on the frontier of energy development and availability."

The following year Berg retired as board chair, succeeded by Centeno. The second unit at the Deely plant became operational, and CPS built a railcar maintenance shop for the 800 cars it owned to transport all that coal from Wyoming.

Environmental concerns and soaring construction costs fueled the debate about whether San Antonio should remain a partner in the South Texas Project. CPS was also aware that the 400 tons of powdery ash produced every day at the Deely coal plant needed a better disposal plan. After searching for options, the company made a deal with Concrete Industries to purchase the fly ash, and the Australian-based company agreed to build a processing plant near Calaveras Lake. CPS's strategy for every aspect of its coal usage was working until the Burlington Northern Railroad began raising its shipping rates despite a contract.

As the 1980s began, coal-hauling rates continued to rise. Cockrell took the offensive, accusing the Department of Transportation and the Interstate Commerce Department of taking advantage of the city. She contacted President Carter, Secretary of Energy Charles Duncan, and the media with her message that the two departments appeared to be saying, "This looks like a good opportunity. These people are now

When Henry Cisneros was elected mayor of San Antonio in 1981, he became the newest member of the CPS board.

using coal, so we'll take advantage of these 'captive' shippers to strengthen the railroads." Once again the feisty mayor urged the City to sue. General Manager Spruce was Cockrell's partner in what she called the last battle in the energy wars. San Antonio successfully sued the railroad and received a rebate and damages, which were directed to citizens through refunds that appeared on CPS bills along with a much-appreciated rate reduction.

By 1981 construction in San Antonio was increasing. CPS installed 73,837 feet of gas mains and 87,574 feet of underground electrical cable in January and February alone. That spring thirty-three-year-old Henry Cisneros became the first Latino mayor of a major city. With degrees from Texas A&M, Harvard, and George Washington Universities, he was also the youngest mayor elected in San Antonio. He praised Cockrell for positioning the city at the forefront of the national trend to diversify energy sources and promised to continue her vigilance over railroad freight rates. He also vowed to search for more competitive natural gas suppliers, tighten fiscal controls on the South Texas Project, and pursue the possibility of using lignite coal, which was in ample supply in Texas, as a fuel source.

In an early press conference Cisneros raised the idea of looking at more innovative fuel sources, including solid waste and solar energy. When Cockrell stepped down as mayor, she became an appointed trustee on the CPS board, where she would continue her commitment to San Antonio's citizens and their access to energy. The five trustees—Chairman Glenn Biggs, a banker; Vice Chairman Ruben Escobedo, a CPA; attorney Earl Hill; Cockrell; and Mayor Cisneros, ex-officio board member—were confident that the energy wars were for the most part behind them. San Antonio was on the rise once again.

As the 1980s got under way, CPS board members were (left to right) Lila Cockrell, Earl Hill, Ruben Escobedo, Glenn Biggs, and Mayor Henry Cisneros (not pictured).

PART 7

The Golden Goose
San Antonio, 1982–1999

In spring 1982 the United States celebrated its Centennial of Electricity. A wave of appreciation for Thomas Edison's breakthrough research in his New Jersey laboratory in 1882 dominated the news and swept through schools and boardrooms. There was no doubt that the power industry had undergone an extraordinary evolution. From those first dynamos, described as "an endless web of belts, clutch pullers, and tubular boilers that could produce 500 volts of direct current" to the massive power plants fueled by gas, coal, lignite, and uranium, electricity had changed the world in remarkable ways.

That year San Antonio celebrated forty years of ownership of its utility company, and CPS was justifiably proud of its fuel diversification program and reliable delivery of electricity to the city.

But at 11:45 on the morning of June 13 an explosion shook the control room of the Braunig plant. Operators Jerome Reininger, Juan Ramirez, and Gary Wickwire were watching Unit 3 come on line when they underwent what they later

An explosion in the Braunig plant's turbine generator room on June 13, 1982, was a reminder that power is not without its dangers.

described as the worst experience of their lives. When Reininger heard the explosion he rushed into the turbine generator room to look for personnel in danger. No one was in the room, but he saw a huge ball of fire between the turbine and the generator.

"I knew I only had a few seconds to take care of some very important business," he said. "The ball of fire was oil vapor, and the noise and vibration and flash all added up to a 'thumbs down' situation and a quick venting of the hydrogen gas. I gave the signal." That signal is a visual command for a complete shutdown of a power unit. When Ramirez and Wickwire received the cue they sprang into action, tripping the unit. Ramirez compared it to being in Vietnam, surrounded by noise, scared faces, a loud rumble, ringing alarms, and bright lights. The cooling system had to be diverted, and every auxiliary system had to be securely turned off. The failure's cause was unknown at the time, but it was soon obvious that damage had been caused by a huge broken blade, or "bucket," from the turbine. Wickwire described the turbine room as "a scene out of a Steven Spielberg movie, with bolts, nuts, oil, door handles, and other equipment all over the floor." Machinists worked in two shifts to repair the unit, and engineers from General Electric, the unit's manufacturer, were sent to San Antonio to investigate. Subsequent stories in the company's in-house magazine and the media served as powerful reminders of the dangers that went hand-in-hand with the phenomenal gift of electricity.

In September one of the pioneers of CPS's fuel diversification program, Victor Braunig, died at the age of ninety-two. In a message to employees Spruce praised the work of his mentor. He noted that 1982 was a "year of anniversaries" and told employees "CPS has marked several major milestones: one hundred years of dependable electric service to the Alamo City, forty years as a municipally owned utility, providing more than $1.2 million in benefits to the City, and sixty years of publishing the *Broadcaster*, the longest running employee publication in the San Antonio area."

As the year ended the company started a Winter Assistance Relief Mobilization program (WARM), recognizing that "feeling warm should not be a privilege; it is a necessity." Designed to provide assistance to citizens with the greatest need, priority was given to low-income, elderly, and disabled customers and those with small children. The program was immediately successful, boosted by the holiday spirit, and over the next decade it would provide more than $2 million in assistance.

There had been a few setbacks, of course. The nuclear plant had encountered construction snags and continued public controversy. The 1979 meltdown of a reactor at the Three Mile Island nuclear plant in Pennsylvania was still on people's minds, and supporters of the South Texas Project found themselves in a lonely spot. Cockrell recalled that it was difficult to convince citizens that nuclear was a good choice and would be the cheapest energy source once the plant was operating. "But we just had to 'suck it up' and stay the course," she said.

"The South Texas Project looked like a disaster at the time," Escobedo said. "Brown & Root had never built a nuclear plant, and there were all kinds of delays until we released them and hired the Bechtel company. The board made the hard decision to stick it out, and every time we voted I got lots of irate calls—some from friends—asking if I was sure. Our view was that power is never about now. It is about the future. It was critical to the growth of San Antonio. 'Power planning' means thinking at least ten to fifteen years out."

The future was on the mayor's mind as well. Cisneros established the Target 90 task force to set goals for the next ten years and to present action items to the city council to be accomplished during its 1983–85 term. Energy was among more than 150 topics the task force considered. Seven major goals were outlined for that sector to accomplish by 1990, with shorter-term goals to be met sooner.

Cisneros's plan asked the City to enact major energy conservation measures, complete the South Texas Project, implement recommendations from CPS's existing long-term energy study, expand the program for senior citizen energy price relief, seek new energy transactions between San Antonio and Mexico, establish a resource recovery energy-generating plant, and enact solar projects. It was tall order, but CPS already was working on most of these initiatives.

Thanks to Bechtel's accelerated production schedule, by 1984 the South Texas Project was 60 percent complete and on track to be operational by 1990. Nearly a decade before, CPS had established its energy systems planning department. Joe Fulton, a Texas A&M University graduate who went to work for the department in 1973, had experienced what he called the "decade of the environment." He had seen the controversy over nuclear energy and been part of the team that canceled the two gas-fueled units scheduled for the Sommers plant and instead designed two coal-fired units for the adjacent Deely plant. By 1990 Fulton was director of the new Research and Environmental Planning Division.

Board member Ruben Escobedo served during construction of the controversial South Texas Project.

Trustees Biggs and Escobedo pushed for the creation of a CPS staff position for economic development. "Our economic lifeline needs a big boost," Escobedo told the board. "Having a beautiful community and ample power supplies will make San Antonio very viable in the competition for new business and industry."

San Antonio was experiencing a building boom, especially in the housing industry, and by

Thanks to an accelerated production schedule, Bechtel got construction of the nuclear plant on track, and it was 60 percent complete by 1984.

the end of the year CPS was servicing 702,000 gas meters. The company had outgrown its headquarters and considered relocating to the site that would eventually house the UTSA downtown campus. Instead, a second office building and parking garage were purchased across from headquarters, connected by a glass skybridge that stretched across Navarro Street.

In 1985, after years of struggling with rising freight costs for coal, CPS awarded a twenty-year contract to Western Railroad Properties, a coal-hauling line jointly owned by the Chicago & North Western and the Union Pacific Railroads. Under the contract CPS agreed to pay between $18.78 and $20.34 per ton of coal versus the $27.66 that had been Burlington Northern's rate. Considering that CPS plants were burning approximately 3.5 million tons of coal a year to provide more than 355,000 customers with electricity, the rate reduction was significant, resulting in multimillion-dollar savings for the company. CPS's new general manager, Arthur von Rosenberg, announced the deal, adding that "a little bit of competition makes a world of difference." A lawsuit against Burlington Northern was still under way; eventually its settlement would bring CPS more good news.

In addition to having strong management skills, gained from working in numerous departments since he joined the company in 1959, von Rosenberg was considered an extraordinary mediator. A petroleum engineer with a degree from the University of Texas at Austin, he studied law at St. Mary's University's night school and, in Fulton's words, "was a sky-high genius." Working with the CPS board and the City, he struck another important deal for the utility company in 1985 with the approval of an out-of-court settlement of the South Texas Project's breach of contract suit against Brown & Root and its parent company Halliburton. San Antonio received $210 million—its share of the $750 million cash settlement—and passed it along to its customers.

That summer, when the Fairmont Hotel was moved from the corner of Commerce and Bowie Streets to Alamo Street, adjoining the La Villita district, CPS moved light poles and traffic signals and temporarily disconnected the underground gas lines, and the 1,600-ton, three-story, turn-of-the-century structure slowly made its way to its new location.

Nearby in the San Antonio Arsenal, built in 1859 to furnish arms and ammunition to the frontier forts of Texas, a fast-growing grocery chain prepared to open its new headquarters. Florence Butt made a capital investment of $60 to start her family-run enterprise, established in 1905 as Mrs. C. C. Butt's Staple and Fancy Grocery. When her youngest son, Howard Edward Butt, took over in 1919, the company name changed several times, eventually becoming H-E-B in

Arthur von Rosenberg became general manager of CPS in 1985.

1942, with headquarters in Corpus Christi. The first San Antonio store opened that year, offering customers the comfort of air-conditioning and the first opportunity to buy frozen food, thanks to the city's early development of refrigeration powered by SAPSCo in the early 1920s. In 1971 Butt passed the leadership role to his son Charles, who had grown the company to 148 stores by the time headquarters moved to San Antonio in 1985. H-E-B would eventually include more than 350 stores across Texas and in Mexico and became one of the city's most generous corporate citizens, recognized for its innovations to the industry, its team-spirited culture, and its support of nearly every good cause in San Antonio.

CPS continued its push for safety and attention to environmental regulations, and in summer 1985 Scott Smith was promoted to supervisor of environmental compliance. As the first Union Pacific trains arrived at the Deely plant, with plans well under way for construction of a J. K. Spruce coal-burning unit, Smith's department was gathering data on ways to make the Wyoming coal—considered to be the cleanest in the country—still more environmentally friendly.

CPS was also investigating the possibility of incinerating San Antonio's garbage in a specially designed power plant, which had been proposed in Target 90. The board appointed an eight-member Solid Waste Advisory Committee comprised of experts in the company, including Fulton, Smith, and environmental lawyer Nelson Clare. Community members were committee chair Richard Howe, an engineer and professor at UTSA; urban planner Robert Ashcroft; Ruth Lofgren, a biologist and president of the League of Women Voters; and Victor Miramontes, a financial consultant who played a big role in crafting the Target 90 goals. According to Fulton, the concept sounded good on paper but "was never a great way to produce energy." He explained that the city was running out of landfill for garbage: "Garbage burning really is a big environmental problem, and we were preparing to build a $100 million plant to do it until, luckily, the City was able to acquire additional landfills."

In 1986 a decrease in oil prices to $11 a barrel reverberated across Texas and beyond, causing T. Boone Pickens, chairman of Amarillo-based Mesa Petroleum, the country's largest independent oil and gas production company, to announce that "oilmen will have to hang on by their fingernails." A banking and real estate slump followed, accompanied by a decline in consumer confidence. San Antonio's engineering and construction industries slowed, although the city's northwest quadrant continued to grow. While not quite as bleak as

When the Fairmount Hotel was relocated in 1985, CPS crews moved light poles
and traffic signals and temporarily disconnected underground gas lines.

the Great Depression, these were dark days. Many banks and savings and loan companies in the city closed, and real estate developers went bankrupt.

Interest in the San Antonio Spurs, which had been playing in the HemisFair sports arena since 1973, also slowed, kept alive by a few committed citizens like B. J. "Red" McCombs, who owned the largest share of the basketball team. When the faltering team won the NBA draft lottery in 1987, it chose U.S. Naval Academy star David Robinson, who would need to complete two years of duty in the Navy before he could play. Sports fans had high hopes that Robinson might be a turning point for the Spurs, and business leaders hoped the development of an extraordinary championship team would someday bring glory—and visitors—to the city. HemisFair had sparked a wave of tourism two decades before. Now several new hotels, the colorful Mercado, the River Walk, and a growing conference business were fueling that economic sector. The recently established Economic Development Foundation was seeking businesses and industries that might further grow the city, and a major selling point was a good supply of energy at one of the country's lowest rates.

Businesses recognized that incentive. Over the next few years real estate developers built the luxury housing projects Oakwell Farms, Elm Creek, and the Dominion. Pape-Dawson, an engineering company founded in 1965 by Gus Pape and Eugene Dawson, was awarded contracts for these residential projects, all of which required utility connections. It also worked on industrial projects, including SeaWorld, a major tourist attraction on the city's northwest side, built in 1988. The nearby Hyatt Regency Hill

Clean coal from Wyoming, less expensive than gas in the 1980s, was transported to San Antonio by rail.

The San Antonio Spurs had first pick in the 1987 NBA draft and chose U.S. Naval Academy player David Robinson.

Country Resort hotel followed in 1989, and the South Texas Medical Center was expanding as well. CPS built its biggest substation to date to accommodate the growing electricity needs in that booming part of town.

Cisneros pushed hard for growth in downtown San Antonio and other parts of the city. In 1985 he championed a bold plan to build an arena near the Convention Center, at the edge of downtown and the city's east side. With a "build it and they will come" optimism, supporters were certain that the Alamodome would position San Antonio to attract the U.S. Olympic Festival and other sporting events, as well as concerts and perhaps an NFL football team. Pointing to Houston's Astrodome as a model, sports fans and other supporters saw economic development possibilities. Others felt that the money should be used for improvements to an aging infrastructure, plans for a light-rail system, and other priorities. Still others were concerned about the toxic condition of the site's soil, created by the long-abandoned Alamo Iron Works next door. The debate was intense, and the political maneuvering that eventually won state support, followed by a petition drive to allow San Antonio to call for an election to vote on the use of a portion of unused VIA sales tax to fund the project, is a complex and interesting story, described in wonderful detail in former mayor Nelson Wolff's book *Transforming San Antonio*. Greehey, Valero's CEO, kept the campaign for the Alamodome alive by hosting civic breakfasts several times a week. An election was finally called, and the project carried by four percentage points. Construction on the 65,000-seat, $186 million arena would begin a few years later. As always, CPS met the massive energy needs of San Antonio's newest landmark.

Cisneros and City Manager Lou Fox also promoted construction of a subterranean tunnel for expanded flood control. The San Antonio River tunnel, with an inlet north of downtown near Brackenridge Park and an outlet south of downtown near the old Lone Star Brewery, would pass beneath the Alamo Plaza, the Hilton Palacio del Rio Hotel, La Villita, and Beethoven Hall and would require $111 million in federal and local funds to build. This was another controversial proposal, and Fox recalled that "the media called the idea 'Fox's Folly.' But we got it passed, raised the money, and the U.S. Army Corps of Engineers started building it in 1987. It took ten years, and when a major flood hit San Antonio in 1998 and the tunnel successfully protected the city, everyone called it brilliant."

At the heart of the downtown area protected by the tunnel, the historic Majestic Theater was beginning to crumble. In 1986 arts patron Joci Straus was asked to organize a group of private

Spurs star Sean Elliott (center), Henry Cisneros, and Lila Cockrell participated in the 1990 groundbreaking for the Alamodome.

citizens who could raise the money to save the building, built in 1928. The slightly newer Empire Theater around the corner was also in need of major repairs. Both theaters reflected the ornate glamour of San Antonio's past and had the potential to become beautiful venues for a growing arts community. Straus established the Las Casas Foundation and led fundraising efforts that produced the $4.5 million to begin restoration of the Majestic. The Empire would wait for nearly a decade, but it too would be restored before the twentieth century's end, thanks to Straus's continued work and a $1 million lead gift from businessman and Spurs owner McCombs and his wife, Charline.

Meanwhile von Rosenberg's negotiating expertise was called into play again, and CPS accepted a $111.5 million offer from the Burlington Northern and Southern Pacific Railroads to settle the lengthy dispute over the exorbitant fees the railroads had charged to transport coal from Wyoming to San Antonio. The settlement's benefits were passed along to the utility's customers through refunds on bills.

Later that year CPS provided electric service for an enormous papal mass in Bexar County. On September 13, 1987, more than 300,000 worshipers heard Pope John Paul II speak from a podium in a huge field, with excellent sound quality and lighting. The pope's message encouraged harmony, compassion, and community service.

Early the next year CPS trustee Cockrell urged the company to establish a community service program, in addition to Project WARM and continued support of United Way. The program became a reality in 1989, and over the years CPS has partnered with more than fifty agencies with beneficiaries ranging from children to seniors.

When Pope John Paul II delivered an outdoor mass to 300,000 worshipers in 1987, CPS provided the power.

Otto Sommers died in 1989 after a long illness, at the age of eighty-three. CPS employees mourned the loss of their former general manager, whose forty-two-year career had spanned so many changes in the energy world.

In the December *Broadcaster* von Rosenberg shared some of the company's recent successes and outlined challenges ahead. He noted that the nation had seen the Berlin Wall come down; closer to home, basketball star David Robinson was playing for the San Antonio Spurs. Von Rosenberg called 1989 a "landmark year in CPS's fuel diversification program" and reminded readers that the coal-fired Spruce unit was under construction and that the second unit at the South Texas Project nuclear plant had gone on line in June, ahead of schedule. The electricity received from the nuclear plant would help maintain competitive rates for customers, and von Rosenberg predicted that the average monthly residential gas and electricity bill in 1990 would be around $69, compared to $75 in 1982.

He congratulated the CPS employees on their record-breaking $380,413 contribution to the annual United Way campaign. "I am looking forward to another challenging year in which we at CPS serve our community not only through voluntarism and the United Way," he concluded, "but also through what will always remain our primary objective: to provide adequate, reliable, and cost-efficient gas and electric service."

A challenge to that objective came in early 1990, when Congress introduced changes to the Clean Air Act of 1977. Fulton told von Rosenberg and the trustees that "of all the bills introduced over the last ten to fifteen years, this one has the biggest potential for having a negative financial impact on CPS." He was quick to say that CPS wholeheartedly supported the reduction of sulfur dioxide emissions by 10 million tons a year by 2001, and the company already had a program in place to achieve that goal. What concerned him was a section of the bill dealing with acid rain and sulfur dioxide—in particular, a subtle change in the definition of a new coal plant.

"A new unit was defined as anything that began operating after the new law took effect," he explained. He told the board that if Spruce Unit 1, already permitted, designed, and under construction, was reclassified this way it would cost customers an additional $100 million to complete. "That was my introduction to government," he recalled, laughing. He traveled to Washington and worked closely with a firm in Georgetown specializing in energy law. "There were about six or seven other utility companies around the country with the same problem," he said, "and we banded together, called ourselves the Class of 1990, and eventually won an exception for those units.

"I gained great insights about how laws get made," he added. "I'd fly back and forth, sometimes more than once a week. I'd get home and have a message that told me to come right back. I'd pick up my mail, put it on the table, and then forget about it as I flew back to Washington. I came home one night and my garage door wouldn't open with the clicker. I went to the front door and there was a notice from CPS that my electricity had been turned off. It was a Friday, and I had to go to the main office downtown and stand in line with about three hundred people. When my turn came I paid my bill, and the clerk laughed when she saw that I was an executive at the company. I was the subject of lots of jokes for a while, but we did get the special allowance for our unit at the Spruce plant."

Unlike natural gas, which produces few by-products when burned, even the highest-grade coal must be constantly monitored to meet standards set by the Environmental Protection Agency, the Texas Air Control Board, and the Texas Department of Water Resources. In response to the changing standards, CPS had established an Environmental Laboratory in the early 1980s, housed in a small red brick building at the edge of Braunig Lake. Its team of engineers and environmental experts had been doing its best to keep pace with the regulations that would be a part of the future of energy.

Part of that future meant replacing the outdated computers in the Gas and Electric Operations Center with in-house software and hardware. CPS also built several large switching stations and substations to keep up with growing power needs in Floresville and the LaVernia area,

Switching stations and substations expanded CPS's power network as far as Floresville and La Vernia.

as well as the electrical needs of companies under construction, including the Golden Aluminum rolling mill plant near Braunig Lake.

When Nelson Wolff was elected mayor in 1991, he was sensitive to the nation's interest in environmental issues and worked with CPS and the city council to continue Target 90's clean energy and conservation goals. He was also a strong supporter of economic development. He grew up on the city's south side, earned a law degree from St. Mary's University, and established the Sun Harvest Farms natural grocery chain. He had also served in the Texas House of Representatives and the Texas Senate and on the city council. During Wolff's two terms as mayor, San Antonio saw completion of the Alamodome, establishment of NAFTA, construction of a new downtown public library, expansion of the San Antonio airport, and consolidation of three water agencies to form the San Antonio Water System. Complicated redistricting issues, contentions on the city council, and a new policy requiring that the CPS board have one member from each of the city's quadrants required sensitivity and consensus building. Wolff was skilled at both.

Early in his term, H.B. Zachry workers cleaned the inlet of the Spruce plant in preparation for flooding the canal. During the summer as many as 360,000 gallons of water a minute would be pumped from the inlet through the plant's condensers. The testing phase of the $444 million plant's coal yard was well under way, a stunning example of CPS's responsiveness to the need for fuel diversification in the production of electricity. Remarkably, CPS paid for its new plant in cash.

In 1993 communications giant Southwestern Bell Corporation relocated its headquarters from St. Louis and, like CPS, H-E-B, and Valero,

When Nelson Wolff was elected mayor in 1991, he became a member of the CPS board and a champion of economic development and clean energy.

became a generous corporate citizen, supporting the arts, education, and economic development. Its first headquarters, at the corner of Broadway and Hildebrand Avenue, required huge amounts of power. CPS was pleased to add the important customer to its load.

The Alamodome opened for business in May 1993, under budget and on time. Cisneros, now U.S. secretary of housing and urban development, accompanied by Texas governor Ann Richards, came back for the grand celebration. The U.S. Olympic Festival was the dome's first event, and the San Antonio Spurs played their first game

The immense Spruce plant cost $444 million, paid for in cash by CPS.

Gov. Ann Richards and Henry Cisneros, secretary of housing and urban development, attended the Alamodome's grand opening in 1993.

in the new arena that fall. A few years later the sports arena that had been so controversial would be deemed too small to attract the football team it hoped for—and also problematic for basketball sightlines.

A controversial idea for another arena began to simmer. Wolff's book provides details about several years of delicate and not so delicate political maneuvers, innovative funding initiatives, changes in Spurs ownership, and management of

a project that would become the new home for the San Antonio Spurs and the San Antonio Stock Show and Rodeo.

In 1993, after five years of operation, the South Texas Project shut down both units for a month to solve problems with its steam-driven auxiliary feedwater pumps. CPS had become less confident over the years in Houston Lighting & Power's ability to act as the project's managing partner, and its concerns increased with the shutdown. Of more concern to San Antonians was Washington's decision in the mid-1990s to close many military bases around the country. A military center for more than a hundred years, San Antonio was home to five bases that were a major part of the local economy.

When plans were announced to close Kelly Air Force Base, the city's largest military employer, San Antonio's leaders worried about the loss of jobs and infrastructure and scrambled for creative ideas. The world-class runway, hangar and warehouse space, inexpensive electricity, and easy rail access could attract commercial aviation companies, and the Clinton administration agreed to give the base a six-year staged closure to allow the city time to bring new companies in. Unfortunately, after decades of industrial activity, the shallow groundwater beneath the base was contaminated and plumes impacted neighboring homes and businesses. The cleanup was extensive and expensive. Liquid wastes were pumped and treated, soil was removed, and landfills were dug up. Once the environmental restoration was complete, the reinvention of Kelly moved forward, and a few years later Boeing brought its first military transport aircraft—a C-17 Globemaster III—to San Antonio for servicing.

In 1996 the CPS board, chaired by Arthur Emerson, approved a $225 million settlement with Houston Lighting & Power to resolve claims related to cost overruns and the 1993–94 shutdown of the South Texas Project. CPS received $75 million in cash and $150 million in savings over the ten-year life of the agreement, and a jointly owned company was to be formed to run the nuclear plant. In the single largest refund in CPS history, $64 million in checks and account credits from the settlement's cash portion were mailed to customers on September 11.

Peter and Julianna Hawn Holt, new majority owners of the Spurs, announced that they wanted to win the 1996 NBA championship and build a new home for the team. Their hopes were shattered when starter Robinson was injured and the team began losing. After eighteen games Spurs general manager Gregg Popovich became the coach; by the end of the season the team was ranked fourth worst in the league. The Spurs won first pick in the NBA's annual lottery and signed a young superstar from Wake Forest named Tim Duncan. While Popovich worked on building a championship team, Holt focused on convincing community leaders to build a new arena. When the Spurs won the 1999 NBA championship, Holt parlayed that success to secure support from a powerful group of business leaders and city and county officials. Wolff's book tells the story, explaining how financial support came from a creative Bexar County bond issue and from the Southwestern Bell Corporation, which received naming rights. The SBC Center (later the AT&T Center) would open its doors on the east side in 2002, with a tremendous electricity load that included remarkable sound and lighting.

In his annual letter to CPS employees, von Rosenberg reflected on the company's satisfaction with the Houston Lighting & Power settlement and what it meant to customers. He announced that CPS's 1997 budget would total $737 million, with an operating budget of $475 million. This would again provide important funding for the city's operations and growth. He warned his team that there was growing interest among state legislators in deregulation and that if the utility industry was opened to private competitors, CPS would find itself in a different business environment. In June the legislative session ended without passage of such a bill, but CPS created

CPS restoration crews worked day and night after a massive storm, with winds clocked at 120 miles an hour, disrupted 50 percent of the power system on May 27, 1998.

a strategic initiatives program so the company would be prepared if the environment changed.

No one was prepared a few weeks before the legislature adjourned when the worst storm in thirty years hit San Antonio on May 27, 1998. It disrupted nearly 50 percent of the CPS system, impacting an estimated 175,000 customers. Winds of more than 120 miles an hour splintered seventy-foot utility poles, and 300 CPS employees were put on duty, with shifts covering twenty-four hours for the two days it took to restore power. Personnel answered telephone calls around the clock, and workers in the field received praise from customers for their courtesy and empathy in a difficult situation for weeks after the ordeal ended.

The team effort was impressive, and von Rosenberg often cited it as an example of what made CPS's customer service special. It was an aspect of the company culture that would be immensely important should the energy market open to competitors. In 1999 he wrote, "As I approach my retirement from City Public Service, I am particularly thankful for the competent and dedicated employees who have made all of our success possible."

In February of that year Jamie A. Rochelle became the first woman general manager at CPS. She had worked for the company for thirty-nine years, following the "analyst line of progression" through the ranks, and held a math degree from

Texas Tech University and a master's in civil engineering from UTSA. She recognized that her promotion marked a change in a male-dominated industry and understood the responsibility. She echoed von Rosenberg's message that major changes were in store for the industry and explained that "employees need to allow themselves to become part of the team. We cannot afford for any one of us to be a passive participant sleepwalking through this transition. Everyone needs to be looking for better ways for getting jobs done."

That fall when the San Antonio Spurs won their first NBA championship, Peter Holt chaired the United Way's general campaign. As a special thank you to CPS employees for their combined pledges of $520,556, he and Spurs player Malik Rose joined Rochelle and the team she was building for a celebratory dinner. There the announcement was made that the combined-cycle power plant under construction adjacent to the Braunig plant would be named in honor of Arthur von Rosenberg.

Conventional power plants convert 35 percent of their produced heat into electricity, but new technology would allow the von Rosenberg plant to convert 50 percent of its heat into electricity. In addition, the plant was already proving less expensive to build than a coal-fired unit. It was also environmentally advanced, reducing nitrogen oxide emissions at record levels. Fulton explained that "burning naturally clean gas fuel in the lowest emitting combustion turbines available, plus utilizing our advanced environment control technology, the new AvR will have the lowest air emission rate of any plant in Texas." Expected to be operational by 2002, the plant exemplified CPS's march into a future with a new emphasis on the environment.

Earlier in 1999 the CPS board had approved a contract to build the Southgate Pipeline, awarding the $43.9 million project to Haines Construction. Phase I would fuel the von Rosenberg unit and was expected to be completed by the year's end. At the time, CPS was purchasing all of its gas from Pacific Gas & Electric, which had acquired Valero's natural gas operations in 1997; it would be delivered through the new pipeline. The Phase II extension, already in the design stage, would create a gateway to the South Texas Pipeline, owned by Houston Pipeline, giving CPS important access to gas suppliers.

Customer access to new electricity providers seemed certain as the twentieth century drew to a close. The *Broadcaster* reported that "electricity has become a commodity, and with deregulation comes choice of provider. In Texas, consumers will decide in 2002 whether they purchase 'private power' from a privately owned company, 'public power' from a publicly owned utility, or 'merchant power' from a nonutility source."

Effective September 1, 1999, Senate Bill 7 gave customers of investor-owned utilities the right to select their energy provider by 2002. Municipally owned utilities and rural electric cooperatives, however, would not automatically participate. Rochelle noted that if the city council exercised its right to opt in to retail choice, outside private entities in the near future could offer electricity to current CPS customers. Deregulation would also allow CPS to provide electricity to customers outside its existing service area. She warned that both scenarios would impact energy reliability and customer service, which had long been the bedrock of CPS.

National and international concerns on the eve of the millennium added to the concern,

With the NBA championship trophy in tow, Peter Holt (far left) and Spurs player Malik Rose (third from left) demonstrated their support for CPS at the utility's United Way kickoff breakfast in 1999. Joining them at the podium were (left to right) Jamie Rochelle, general manager of CPS, and Anthony Edwards, Sharon Luther-Minor, Angie Huntington, and Sara Lance.

specifically that businesses, including airlines and utilities, might be vulnerable due to flaws in older software, with airplanes colliding, massive systems crashes, and a collapsing power grid. The twentieth century ended with this message from Rochelle:

> The new millennium is fast approaching, and CPS has been Year 2000 ready

since July. Hundreds of CPS employees will remain on call or on the job covering contingency plans while the spectacular celebration rolls through each time zone around the world. . . . Our employees' dedication to the community, along with San Antonio's support of its city-owned gas and electric company, are priceless traditions that must be preserved.

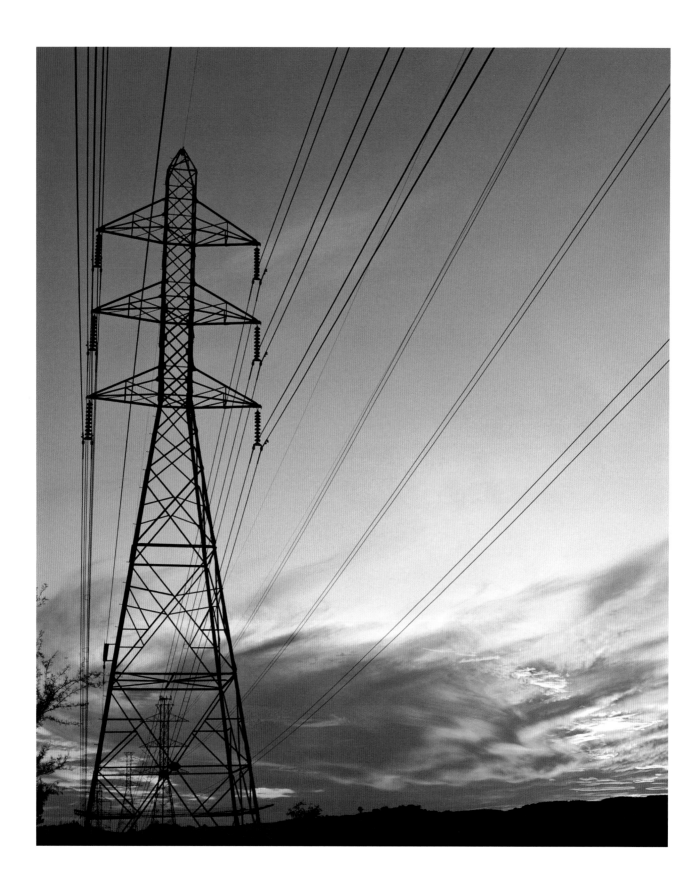

PART 8

Into the New Millennium
San Antonio, 2000–Present

Since those last days of the twentieth century, nearly two decades have brought extraordinary advances to nearly every aspect of modern life.

"No one can foretell the future exactly," Rochelle said, in her State of the Utility Address in early 2000 at Villita Assembly Hall, "but the experiences and observations of the people at CPS are a critical link to the utility's future." She urged the company to solidify its collective vision in preparation for what lay ahead and congratulated employees on the company's 1999 Texas Environmental Vision Award for outstanding recycling efforts.

CPS strengthened its game plan in response to Rochelle's call to be ready for the changing energy landscape, and the next seventeen years would see a spectacular boom in the development of renewable energy. The company would eventually become the country's largest municipally owned purchaser of wind power and a giant presence in the solar energy field.

The world did not end, as some had predicted in 1999, and citizens turned their attention to the new millennium.

San Antonio was now the seventh largest city in the United States. In 2000 its international airport included twelve airlines, offered 129 daily departures to twenty-nine destinations, and handled 3.6 million passengers. Over the next decade the city's and county's aggressive economic development would attract new businesses, including the PGA Village, a Toyota Tundra plant, Rackspace, Amazon, Microsoft, and Google Fiber. A major factor in their selection of San Antonio would be the municipally owned utility's ability to deliver reliable electricity at reasonable rates.

In late 2001 Valero, the prize the city had received more than two decades earlier in its settlement with Coastal States, planned to build a new headquarters on a 148-acre site in northwest San Antonio. Valero had supplied natural gas to the city until 1997, when it sold its gas operations to Pacific Gas & Electric. Valero had steadily shifted its focus to oil refineries and pipelines, and over the next decade it would expand to include sixteen refineries in the United States, Canada, the United Kingdom, and the Caribbean; eleven ethanol plants; and a fifty-megawatt wind farm. As its acquisitions accelerated, Valero purchased a California refinery owned by ExxonMobil, as well as Exxon-branded service stations, and began retailing gasoline under the Valero brand.

Greehey had been the company's CEO since its start. Known for his tenacity and can-do spirit, he had seen Valero's workforce increase to 20,000 and its assets rise to more than $40 billion. He grew up in Fort Dodge, Iowa, and was working after school to help support his family by the time he was twelve. He joined the Air Force to take advantage of the GI Bill. Stationed in San Antonio, he graduated with honors from St. Mary's University in two and a half years and

Valero Energy, the company that came to San Antonio as a result of the lawsuit between the City of San Antonio and Coastal States, grew into a Fortune 500 corporation with assets of more than $40 billion.

was the first in his family to attend college. While he earned his degree in business administration, he worked nights and weekends parking cars at a hospital garage and balanced a young family. He became a CPA with Price Waterhouse, later worked as an auditor for Exxon, was hired by Coastal States in 1963, and soon became its senior vice president.

When Coastal States divested itself of LoVaca, it was the largest spin-off in U.S. history. Greehey felt a connection to San Antonio and recognized that the new company, which became Valero in 1980, could have a big impact on the city's future. Its first years were difficult, however, as it had debt of $700 million, and the plunge in gas prices in the 1980s was an added blow. Greehey's decision to spin off a substantial portion of the gas pipeline operation and focus on refineries and fuel oil paid off, and by 1991 Valero was in the black. A few years later it realized its first big profits, and by 2002 it was the largest independent oil refiner in the country.

The partnership forced on Valero and CPS through the legal settlement blossomed over the years, and the two companies shared many corporate values. Both built employee-oriented cultures, considering their workforce as their most important asset. Neither ever laid anyone off. And from the beginning, both incorporated giving to the community into their company philosophies.

Greehey served on the boards of the United Way, Southwest Foundation for Biomedical Research (now the Texas Biomedical Institute), Cancer Therapy and Research Center, and many other nonprofits. In 2002 Valero took over title sponsorship of the Texas Open, the country's third oldest PGA tour tournament, first played in 1922 at the Brackenridge golf course. As the Valero Texas Open, it quickly became the annual leader in charitable fundraising, raising more than $100 million for children's charities over the next two decades.

Children were a major focus of CPS's philanthropy as well. From Tuttle's first requests in 1922 that employees support United Way, the utility understood its good neighbor relationship with the citizens of San Antonio. It established a long list of annual giving campaigns and sponsored events, including its annual Kids Fish Day at Calaveras Lake, where children enjoyed outdoor activities with CPS volunteers.

Kids Fish Day at Calaveras Lake is part of CPS Energy's ongoing involvement with customers, employees, and the community.

San Antonio's population had grown to more than 1.2 million, and between 2000 and 2005 it was the fastest growing of the ten largest cities in the country. Additional improvements to the international airport and the University Health Care System were under way, and biomedical and technology industries began to boom. The old Beckmann quarry in northwest San Antonio would be developed into La Cantera, a shopping center, resort hotel, and golf course. The Rim surrounding it would include housing, restaurants, shops, theaters, offices, and a hospital.

Downtown saw transformation as well, with financial support coming from corporate citizens AT&T, H-E-B, and Valero. With 14 percent of its gross revenues going to the city operating budget each year, CPS Energy helped San Antonio become a generous participant in public-private partnerships. Today the Tobin Center for the Performing Arts, incorporating the historic Municipal Auditorium, brings a variety of music, dance, and theater events to the community. The City, Bexar County, and the private sector

Horizontal directional drilling was used in 2000 to install pipe beneath freeways, the San Antonio River, and Braunig and Calaveras Lakes.

collaborated on an ambitious expansion of the San Antonio River, anchored by the multi-use development of the Pearl Brewery to the north of its two-mile tourist hub and the historic missions to the south. The thirteen-mile linear park includes walking and bicycle trails, picnic areas, and public art and will eventually have access to the EPIcenter multi-use complex. Envisioned as a national energy think tank and demonstration facility, the EPIcenter's location in the renovated Mission Road plant, built nearly one hundred years ago, seems appropriate.

Well-run operations and the preparedness thinking that had been a part of CPS Energy's culture since its earliest days had positioned the company for the energy needs of the new millennium. In 2000 CPS purchased the independent utility system at Kelly Air Force Base as part of a federal requirement of the 1996 Base Realignment and Closure Program (BRAC). After examining the system, CPS found that the utilities were not up to date, and major changes were required to integrate it into the system. The company went to work installing meters, streetlights, and fourteen transformers for the former base's one-million-square-foot hangar. Purchased for $14.5 million, with $6 million up front and the rest paid over twenty years, depending on the revenue stream, Kelly became the nation's largest military utility system to be privatized. Major tenants included Boeing, Pratt & Whitney, General Electric, and Lockheed Martin. The City began negotiating with the Air Force on similar projects planned for its Brooks, Lackland, and Randolph bases and at Fort Sam Houston.

CPS employees Russell Anderson (left) and Arthur Randall (right) worked in the Salado Street Garage when it was expanded in 2000 to accommodate CPS's inventory of construction machinery and more than 3,000 vehicles.

Sixty miles of the Southgate pipeline had been completed with horizontal directional drilling technology that allowed pipe to be installed under Calaveras and Braunig Lakes, freeways, and the San Antonio River. Rigs were housed in the 48,400-square-foot Salado Street Garage that had once been the site of the gas plant Tuttle helped build in 1907 as an employee of American Light & Traction. The fleet operations department had almost doubled its inventory since 1998, from 1,800 to 3,000 vehicles, and the garage was a bustling hub for maintenance of rock saws, drilling rigs, and other equipment and repairs such as retrofitting trucks with tool mounts, voltage power sources, and lighting.

Before most of this incredible growth took place, however, there was a wild spike in gas prices coupled with the state's coldest winter in 106 years. Rochelle reported that 2001 had been difficult for the industry. Problems in California that included rolling brownouts shook the western grid and forced California utilities into bankruptcy. Closer to home, it was so cold in San Antonio that swimming pools froze. Of course, customers saw a jump in their utility bills and were upset, as they had been during the energy crisis of the 1970s. In response, CPS established its Residential Energy Assistance Partnership (REAP) in 2002, which provides power to the community's disadvantaged, disabled, and elderly populations through donations that customers make when paying their energy bills.

"It's clear that we do not have a national plan for restructuring power," Rochelle reported in her year-end message, "and our nation does not yet have sufficient infrastructure to support an open retail electric market." She thanked employees for "lending a helping hand to weatherize homes, donate blood, collect and distribute blankets in

As the twenty-first century began, CPS modernized its hiring practices. Yolanda Olivarri was one of the company's first female utility workers.

winter and fans in summer, for giving to United Way, and for keeping our Texas grid reliable."

Rochelle also reported that CPS was in step with the national trend to open employment opportunities to women. For decades many of the company's jobs had been perceived as best filled by men because of the strength requirements and risks involved. As that perception changed, the company saw women utility workers, truck drivers, foremen, and of course, the CEO.

The terror attacks in New York, Washington, D.C., and Pennsylvania on September 11, 2001, rocked the nation and the world, and Rochelle thanked her team for "staying calm and effective on that terrible September day, reassuring and responding to customers despite the tragedy, and for adapting to the new security measures that we've taken to make our facilities safer."

A state law, passed in 1999, required utilities in Texas to begin renewable energy programs, and Rochelle had been fielding questions from employees about whether renewables would harm electric sales from existing plants. "Think of our customers," she said. "Most do not know much about electricity production, but they do have strongly held values about protecting the earth and finding advanced ways to access energy and save money." Relying on veteran employees like

Fulton and Smith, the company began its push to be a leader in wind-generated electricity, solar projects, and fuel cell development by the end of the decade.

CPS made its commitment tangible when it introduced its first renewable product, Windtricity, in 2000. The program urged customers to purchase wind power at a slightly higher cost than traditional electricity, knowing that "every kilowatt of wind you purchase supports the continued development of renewable power." The Desert Sky Wind Farm, through which CPS contracted for wind power from American Electric Power, was completed in 2002. Gov. Rick Perry officiated at the dedication, joining utility company executives and the media in the small West

Wind power became part of CPS's portfolio of energy in 2000.

Texas town of Iraan where 107 wind turbines dominated a bright blue sky, marking the arrival of wind power in the state.

The CPS board fully supported the strong move toward renewable energy. Longtime trustee and board chair Wolff had been a big proponent of renewables from his early years on the city council and as mayor. When he retired from the board to run for Bexar County judge, and won, he continued to work closely with the company to bring business opportunities—and power customers—to San Antonio. Cheryl Garcia filled Wolff's vacant position, representing the city's northwest quadrant. Stephen S. Hennigan, then CFO of the San Antonio Credit Union, represented the

northeast quadrant. Clayton T. Gay Jr., a CPA from the southeast quadrant; Alvaro "Al" Sanchez Jr., chief of base realignment at Kelly, from the southwestern quadrant; and Mayor Edward Garza, who replaced outgoing mayor Howard Peak, completed the first CPS board of the new millennium.

In 2002 most of the trustees joined other city leaders and officials and a crowd of sports fans when the SBC Center opened with appropriate fanfare. Julianna Hawn Holt had selected colors and materials for the arena's dramatic interior and, along with art consultant Alice Carrington, had chosen artworks created by local artists. George Cisneros's thirty-one-foot-long *Atomic*

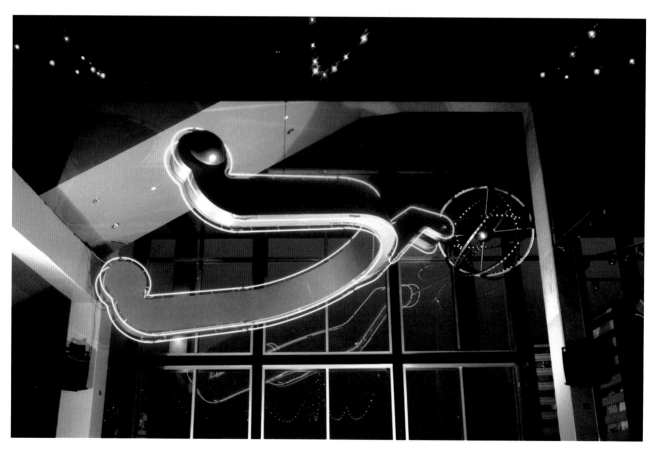

The SBC Center (renamed the AT&T Center), with extraordinary high-tech lighting, became the San Antonio Spurs' home when it opened in 2002.

Spur, illuminated by 500 metal halide lights in the arena's ceiling, captured the flash and excitement of the San Antonio Spurs' new home. They would win their second NBA championship in 2003, fulfilling San Antonio's wildest dreams for its home team.

Rochelle retired as general manager in 2002, succeeded by Milton B. Lee. Hired at CPS as senior vice president of electricity transmission and distribution in 2000, he had been the general manager of Austin Energy for nearly a decade. Lee, with a degree in mechanical engineering from the University of Texas at Austin, had worked as a junior engineer at General Electric and in positions at the Lower Colorado River Authority and Austin Energy. He was the company's first African American general manager and also its first leader not to come through the company's ranks.

At first some employees were disappointed that Lee had not spent decades working at CPS, as all of its previous leaders had. Rochelle assured employees and trustees that she had observed his management skills for more than twenty years and was confident of his expertise in planning, constructing, and operating electrical systems and negotiating acquisitions and transportation of fuels. He was also a leader in renewable energy. "Nothing is broken at CPS," Lee reassured employees. "I don't need to come in and fix anything. But I will be looking to do continuous process improvement." The company recognized that top performance and customer service were essential in an increasingly competitive industry, and Lee promised to lead with that in mind. With San Antonio's population growth of nearly 6 percent between 2000 and 2003, CPS added four natural gas peaking units at the Leon Creek plant in 2004 to meet its peak load.

Milton Lee became general manager of CPS when Jamie Rochelle retired in 2002.

The city's courtship of Toyota began not long after Lee was promoted, and he was quick to tell the Japanese car and truck manufacturer that San Antonio could offer cheap and ample power. In addition, the city had a bountiful, trainable workforce and was close to the Mexican market, and its residents loved pickup trucks. The city, county, and state governments put together a $133 million incentive package, and four years later the plant opened with 2,200 employees building the Toyota Tundra pickup truck model.

CPS acquired Brooks Air Force Base's gas and electric systems in 2003 and those of Fort Sam Houston in 2004. As with the Kelly purchase, repairs and renovation were needed. More than eighty BRAC-related projects were scheduled, and the utility company had to work fast, installing two hundred temporary and permanent electric services and extending existing gas mains. The U.S. military committed $3 billion for construction of the new projects, most of which would be at Fort Sam, with smaller ones planned for the Lackland and Randolph Air Force bases and Camp Bullis. According to projections, the BRAC-related construction, with an estimated completion date of 2011, would boost San Antonio's economy by 15 percent. "The government is not our typical customer," CPS's energy solutions manager told the *Broadcaster*, "but it's good to know that the work we are doing is benefiting our military and our community."

After decades of doing business as City Public Service, in 2005 the company officially changed its name to CPS Energy and adopted a new logo. A new mayor and city manager gave San Antonio a double burst of energy and know-how; their teamwork would result in important ventures in arts and culture, civic engagement, and economic development. Phil Hardberger moved from West Texas to San Antonio with his wife, Linda, in 1970. After a successful career as a trial lawyer, he was appointed to the Fourth Court of Appeals and was later elected its chief justice. When he retired in 2003, San Antonio's business leaders urged him to run for mayor. Reluctant at first, he eventually agreed and was elected with strong bipartisan support, a rarity in city elections. His intelligence, outgoing personality, and political experience positioned him to be a very effective mayor. An early accomplishment was the recruitment of Sheryl Sculley as city manager. Sculley was equally smart and savvy about running a city, having served as the assistant city manager of Phoenix for sixteen years and as city manager of Kalamazoo, Michigan. Hardberger and Sculley set out to update and streamline administrative

systems and to fuel the momentum that would make San Antonio a leading city in the new millennium.

When Hurricane Katrina struck the central Gulf Coast in August, killing nearly 2,000 people, causing more than $108 billion in damages, and displacing more than a million people, nearby cities were essential to restoration and evacuation efforts. CPS responded immediately, assisting Entergy Corporation, the utility serving southeast Texas and Louisiana, by providing personnel and equipment for the immense task of power restoration for millions. Al Lujan, senior vice president of Energy Delivery Services, remembered the hurricane's aftermath as one of the most challenging experiences of his life. It was also "one of the proudest," as he saw his team help a devastated region recover from the tragedy. San Antonio opened its doors to nearly 35,000 evacuees and set up shelters at Kelly Air Force Base, the old Levi Strauss facility, and abandoned parts of Windsor

Bexar County Judge Nelson Wolff, City Councilwoman Patti Radle, Valero CEO William Greehey, Mayor Phil Hardberger, and City Manager Sheryl Sculley (left to right) were forces behind Haven for Hope.

The thirteen-mile expansion and enhancement of the San Antonio River Walk unified neighborhoods and created one of the country's most beautiful linear parks.

Park Mall (which eventually became home to tech giant Rackspace).

Media coverage of makeshift shelters and efforts to provide comfort and services to those who had escaped the wrath of Katrina expanded to include the broader systemic problem of homelessness, which was growing in San Antonio. Greehey and Valero had been generous supporters of agencies that provided shelter and food to the homeless, and Greehey had become convinced that an all-inclusive, holistic approach might create a more successful solution. His vision for the Haven for Hope project became a major public-private partnership, with Hardberger, Sculley, and Wolff as its stalwart champions. It would provide a model for other citywide and countywide projects and would further solidify San Antonio's reputation as a city that was pioneering important change. In 2006 a twenty-two-acre site west of downtown was purchased, and construction of the Haven for Hope campus began. It would eventually accommodate 1,500 homeless people, offering meals, health care, drug and alcohol rehabilitation, education, and vocational training that had

the potential to transform lives. Funding would come from the Greehey Family Foundation and other private donors, as well as from the City and County. More than thirty community partners would join the effort to provide services.

Another ambitious project began around the same time. The brightly lit, two-mile section of the San Antonio River that flowed through the downtown tourist district had long been a symbol of the city's special charm, but longer sections to the north and south were in various stages of neglect. The City, County, San Antonio River Authority, and San Antonio River Foundation, a new private fundraising entity, joined forces to enhance a thirteen-mile stretch extending from the river's headwaters near Brackenridge Park through downtown to the city's southernmost Spanish mission, Espada. AT&T Corporation had completed its acquisition of SBC Communications, and Hardberger convinced the company's dynamic CEO, Ed Whitacre, to champion the project. AT&T's lead gift of $15 million started a decades-long project that would join the city's north, central, and south sides with a walking path, public art, river and habitat restoration, and recreation opportunities.

Meanwhile in West Texas, a few miles from Sweetwater, construction began at the Cottonwood Wind Farm. CPS Energy signed a twenty-year contract with a Houston-based renewable energy developer that owned and operated the farm to purchase all of the wind power produced by the farm's sixty-seven wind turbines. Not far from the Desert Sky Wind Farm, the turbines' additional output gave CPS Energy the largest wind capability of any municipally owned utility in the country.

With the additional output of the Cottonwood Wind Farm in 2006, CPS had the largest wind capability of any municipally owned utility in the country.

Lee had developed a CPS Energy strategic plan that he described as "a bridge between annual operations and long-term goals," featuring wind and solar energies. He assigned oversight of renewables, climate change, and other environmental issues to his deputy general manager, Steve Bartley.

The strategic plan included actions to meet the energy needs of a growing population. CPS Energy had recently acquired an additional 12 percent of the South Texas Project's nuclear plant, and Zachry Construction would build a coal-fired unit on the shores of Calaveras Lake at the J. K. Spruce plant, projected to be operational by 2010. "The last solid-fuel generating plant we built was in 1992," Lee reminded employees and customers. He committed to the community that the new plant would meet or exceed every environmental requirement.

CPS Energy's Environmental Laboratory, dedicated to ensuring that its power plants complied with stringent rules and regulations set by state and federal agencies, would soon be expanded, incorporating sustainable energy features like solar panels and LED lighting. In 2006 CPS Energy built its first baghouse, described as a "giant 115-foot-tall vacuum cleaner" for the fly ash produced at the Deely plant. Boral Technologies partnered with the company to continue recycling the substance for use as a concrete enhancer to help prevent potholes and other damage to new roads and highways; plans were in place to build another baghouse in 2007.

San Antonio was receiving its electricity from more than a dozen generating units and eighty-six substations spread all over the city. Four substations had recently been added to accommodate the power needs of the Southwest Research Institute, the South Texas Medical Center, residential growth in the city's northwest quadrant, and the Toyota vehicle assembly plant.

An even bigger boom occurred in 2007—one of the biggest South Texas has ever experienced—when the Eagle Ford oil/gas/condensate play was discovered. The economic impact was enormous for oil and gas production, jobs, construction, transportation, and local government and state revenues. Towns that had been dormant for years were jolted awake by employment opportunities and money that poured in. Restaurants, bars, motels, and even a Lucky Eagle Casino were built almost overnight, and small towns like Alice, Carrizo Springs, Cotulla, and Eagle Pass became hot spots. Railroad spurs were built to transport equipment and sand to drilling rigs

being constructed in more than fifteen counties; truck sales doubled and tripled, energy loads were immense; and the next eight years brought an incredible boost to the state's economy. CPS Energy would seize the opportunity to negotiate a new natural gas transportation contract resulting in construction of a pipeline connecting with the Southgate pipeline to expand the utility's sources of natural gas.

Lee's dreams of momentum for solar energy were realized in 2008 through a partnership between CPS Energy and Silver Ventures, a forward-thinking company that was developing the Pearl Brewery into a multi-use complex with offices, restaurants, retail, and residential apartments. CPS Energy would be able to test the viability of solar energy in a large commercial application. When the panels were mounted on top of the 67,000-square-foot Full Goods

CPS Energy partnered with Silver Ventures in 2008 to install solar panels for the 67,000-square-foot Full Goods Building in the Pearl Brewery's multi-use development project.

Building, San Antonio had the largest solar installation in the state. The building was also equipped with the most energy-efficient air-conditioning system available.

"We are changing with the times," Bartley said, pointing to development at the Pearl as a concrete example of the Vision 2020 plan CPS Energy developed that year. In addition to environmental issues and renewables, the plan focused on the company's changing workforce, as baby boomers began to retire; an aging power infrastructure; and the emerging global market with competitors worldwide providing transformers, generation equipment, and even utility poles.

Many utility poles and equipment needed to be replaced in July, when Hurricane Dolly slammed into the Texas coast. Wind, rain, and lightning caused extensive damage to the Brownsville electric system, and a thirty-four-member team from CPS Energy spent a week there helping restore power to more than 4,000 customers.

San Antonians had enjoyed stable service for more than a decade, with residential bills maintained at 10 to 15 percent below the best market prices in major Texas cities. Through printed inserts in bills, and on radio and television spots, customers were encouraged to take advantage of cost-saving measures like free Honeywell thermostats, which helped reduce air-conditioning use, as well as the Windtricity program and solar offerings. As part of Vision 2020's goals, CPS Energy announced plans to lower peak demand for electricity by 771 megawatts from the 2008 demand; aim for a renewable energy capacity of more than 1,200 megawatts in its portfolio; and develop a carbon-reduction strategy for slowing the increase in carbon emissions.

Funded by a small rate increase in 2008, a pilot program for an advanced metering infrastructure was developed as part of a multiyear process to replace 700,000 electric meters and 320,000 gas meters with "smart meters." Forty thousand were installed across the city between 2010 and 2012 to test the efficacy of the new system. Ultimately a different meter type was selected for the program, which began in 2014. CPS Energy added another wind farm to its portfolio, this time on the Kenedy Ranch near the Texas coast, and signed a fifteen-year contract to purchase that output. It began work on an expansion of the South Texas Project nuclear plant. The utility had made a strong start toward meeting its 2020 goals, and it earned national recognition as a "top utility" from *Site Selection* magazine for helping to energize economic development.

While they were proud of development in northwest San Antonio, Hardberger and Sculley did not want the city's downtown area to be ignored. Now that Zachry Construction had completed the Museum Reach section of the River Walk, dazzling residents and tourists with its artwork and beautiful walkways, they continued to push for the river extension project, focusing on the long stretch south to the missions. The renovation of Main Plaza, once the heart of San Antonio's business and cultural life, was under way, and plans to redevelop the HemisFair grounds were in the design stage. Several large hotels, including the 1,000-room Grand Hyatt, were built to accommodate the booming conference and tourism industry, one of the city's largest revenue sources.

In 2009 Julian Castro, age thirty-four, became the youngest mayor of a top-fifty city in the United States, bringing youth and vibrancy to the office much like Henry Cisneros had done in the early 1980s. Castro was a double major in political science and communications at Stanford University and had a law degree from Harvard University. He was committed to the projects begun by Hardberger, admired Sculley's management of the city, and was welcomed by the city council, where he had served for two terms as its youngest member.

After the South Texas Project's plan to add two more nuclear units to its plant stalled because of rising costs and continued public controversy, and media attention about difficulties between the city council and the CPS Energy board escalated, Milton Lee retired. Jelynne LeBlanc Burley was appointed acting general manager while the board conducted a national search to find Lee's replacement. Doyle Beneby was hired nearly a year later, taking over the leadership of CPS Energy on August 1, 2010. Beneby had grown up in Miami, Florida, and attended Montana Tech on a basketball scholarship. He returned to Florida, worked for Florida Power & Light in a variety of positions, and earned an MBA from University of Miami. He was hired as general manager of the Consumer Energy Company in Michigan and in 2003 was recruited by the Exelon Power Company in Chicago, where he eventually became president.

Beneby, like Lee, did not come from within CPS Energy, and he spent his first months building employee trust and teamwork. In recent years employees had been asking for more transparency, and Beneby and the board complied. For the first time CPS Energy provided salary information for its top twenty employees, the cost of annual legal and consulting fees for the recently halted expansion of the South Texas Project plant, the cost of the Windtricity program, and names of its top ten customers. The new CEO was impressed by

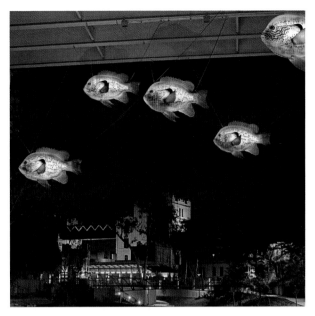

The Museum Reach of the San Antonio River dazzled everyone with its spectacular lighting and public artworks.

When Julian Castro was elected in 2009, he became the youngest mayor of a top-fifty city in the country and a member of the CPS Energy board.

the employees' work ethic and commitment to the community, and he reassured them that the state of the utility was healthy, "with good generation assets, including a low-emission coal plant that was just dedicated. I have a good relationship with the company's trustees; and I speak often with City Manager Sculley, Robert Puente at SAWS, and Keith Parker at VIA. These strong relationships are essential to our future."

Future planning was on the minds of Beneby, CPS Energy trustees, and employees. "The question is, what do we want to be as a company, and what should we be doing to get there?" Beneby said in an early meeting with the board. He was an advocate for renewable energy based on cost effectiveness and pushed a bold vision for solar energy. "No place in the United States is the solar capital yet—a hub for research, development, design, manufacturing, assembling of solar components. I have my sights on San Antonio to do this." He outlined his main goals for the year, starting with development of a larger solar presence, either through building a plant or purchasing more solar agreements. He also vowed to concentrate on managing the company's capital costs to minimize the rate increase that was projected for 2012.

While Beneby was setting goals for CPS Energy, Castro was doing the same for San Antonio through SA2020. Similar to Cisneros's Target 90 nearly two decades before, Castro's vision plan outlined where the city wanted to be in ten years and how to get there.

The housing market fell sharply in 2010 after a decade of booming construction and sales. As construction dried up, truck sales at the three-year-old Toyota plant declined and the company cut production, froze pay, shortened work hours, and laid off many of its temporary employees.

Doyle Beneby assumed the newly established position of president and CEO at CPS in 2010, following Milton Lee's retirement as general manager.

When a major earthquake and tsunami in Japan disrupted its parts supply line in 2011, the company halted production at its North American plants and used the down time to upgrade equipment, including the addition of robotic assembly parts. Production resumed in 2012, and in 2017 a truck is made every 61 seconds.

Not far from the Toyota plant, on the city's southeast side, the fourteen-megawatt Blue Wing solar facility, with 214,500 solar panels situated on 140 acres, was dedicated in 2010. CPS Energy signed a thirty-year contract to purchase 100 percent of its output, officially adding solar power to its portfolio.

Like previous leaders, Beneby became active in the community, serving on the boards of the Chamber of Commerce, the Economic Development Foundation, the United Way, and

The Blue Wing solar facility became operational in summer 2010, expanding CPS Energy's commitment to renewable power.

the San Antonio Medical Foundation. He continued his push to transition the utility company toward lower carbon-intense fuel, clean coal, natural gas, nuclear, and a combination of wind and solar. The company established partnerships with seven "clean tech" companies and signed an agreement with UTSA to conduct energy and renewable research, led by energy policy expert Les Shephard.

As the decade began, the much-anticipated Haven for Hope and Morgan's Wonderland, an amusement park for children with special needs, opened their doors, both reflections of San Antonio's strong community spirit. That spirit was evident at CPS Energy as well, through its Casa Verde program to weatherize the homes of disadvantaged customers. The company provided this service to more than 14,000 homes over the

Helping our Customers

10,000 Homes Weatherized

CPS Energy maintains the good neighbor relationship established in its earliest days by providing assistance to customers in a variety of ways. Its weatherizing project has expanded from 10,000 to more than 14,000 homes.

next six years. The city council also approved CPS Energy's request to fund the Save for Tomorrow Energy Plan, which offered rebates and incentives totaling more than $849 million over the next decade to encourage customers to make energy-efficient improvements.

By 2011 CPS Energy was maintaining more than 700,000 gas meters. The company's transmission lines now stretched 1,474 miles, and there were 12,001 miles of distribution lines—a drastic increase from the 50 miles of lines that had belonged to the utility company when the City purchased it in 1942. Ten major transmission projects had been completed between 2009 and 2011, and more than one hundred substations and switchyards were in operation. The peaking units at the Braunig plant were operational, capable of producing 190 megawatts of quick-start generation. Best of all, in the minds of employees and the board, CPS Energy placed first in the customer satisfaction survey conducted by J.D. Power & Associates.

Moving toward a future of energy conservation, the company made a bold decision to

decommission and close the Deely coal plant in 2018. Sculley considered it a brave, intelligent move, noting that the closure would bring air quality improvements to the city and adding that "the CPS Deely plant is the first municipally owned coal plant announced to retire in Texas."

With the decision to decommission the Deely plant, CPS Energy was hesitant to invest in expansion of their oldest coal-fired unit. Instead they purchased the Rio Nogales plant in Seguin, a gas-powered unit that allowed the company to increase its low-carbon fuels and proactively save as much as $500 million that would have been used for Deely plant improvements. Consistent with its pursuit of cleaner energy, the company installed 120 public charging stations throughout greater San Antonio and partnered with five clean energy companies that had the potential to bring jobs and educational investments to San Antonio.

In 2014 President Barack Obama selected San Antonio's mayor as the U.S. secretary of housing and urban development. When Castro moved to Washington, D.C., city councilwoman Ivy Taylor was selected to complete his term and also took

After CPS Energy decided to retire its oldest coal plant, it purchased the gas-fired Rio Nogales plant in Seguin to ensure delivery of ample energy to customers.

his place on the CPS Energy board. Other board members included Ed Kelley, former president and CEO of USAA Real Estate, serving as board chair and representing the city's northwest quadrant; vice chair Derrick Howard, executive director of the Freeman Coliseum, representing the southeast quadrant; Homer Guevara, an economics professor at Northwest Vista College, representing the southwest quadrant; and Nora Chavez, managing director at Stifel Finance, representing the northeast quadrant.

That year, after analyzing its 2010–2012 pilot program, the company kicked off its five-year, $290 million grid optimization project, designed to install an advanced metering infrastructure that enables automation of electrical distribution. Widespread meter installation began in 2014; by 2016 the number had grown to 226,000, and in

2017 there were 830,000. By 2018 CPS Energy expects to have 1 million smart meters in service.

From the earliest days, when Tuttle told employees of the American Light & Traction–owned utilities in San Antonio that community involvement was part of who they were, the employees listened. Each year contributions to United Way increased, and in 2015 pledges reached $1 million.

Nearly every year since 2005, CPS Energy has earned J.D. Power's highest customer satisfaction award for gas service in the southern region of the country. Tuttle and the company's other early leaders would be proud of the extraordinary philanthropic dollars given to the community and the high level of customer satisfaction. The company's thirteen general managers and CEOs were people who believed in hard work, tenacity, integrity, and compassion, and they established a culture that carried those values forward. The company's policy of caring about its employees has also stayed strong. From its baseball teams and social clubs to the *Broadcaster*'s articles about employees who became war heroes or new mothers, from sharing health tips in the 1930s to its current wellness program with access to a wellness coach, CPS Energy is like a family.

When Beneby retired in 2015, CPS Energy appointed Paula Gold-Williams interim president and CEO. She had worked for the company for eleven years, rising through the ranks to group executive vice president of financial administrative services, CFO, and treasurer. Gold-Williams grew up in San Antonio and earned an associate's degree from San Antonio College, a BBA in accounting from St. Mary's University, and an MBA in finance and accounting from Regis University in Denver.

Kelley chaired the search committee and enlisted an executive search firm to assist in the process. During the yearlong search the committee interviewed several dozen applicants and selected five finalists. "At the end of the day, we knew we had the best candidate already in place," Kelley said. "Paula Gold-Williams won it fair and square." When she became CEO she promoted Cris Eugster, who held an undergraduate degree from Texas A&M University and a doctorate in electrical engineering from the Massachusetts Institute of Technology, to chief operating officer. Gold-Williams hosted employee town hall meetings, expanded solar power, and focused on the community.

In partnership with OCI Solar Power in San Antonio, CPS Energy has come close to making

"In my twelve years with CPS Energy, and my first year as President & CEO, it's my tremendous honor to serve our community. I am fortunate to lead an engineering and analytics firm, one that has a long and rich history. I'm proud that our organization is striving to provide innovative energy solutions and is nimble at keeping our People First philosophy strongly progressive, adding value to the lives of our employees, customers, and community."

Paula Gold-Williams,
President & CEO, CPS Energy

Paula Gold-Williams became CEO of CPS Energy in 2016.

Beneby's vision of a "solar mecca" a reality. The most recent solar facility was completed in 2015 on 900 acres of land near Uvalde. Alamo 5 boasts 9,000 dual-axis trackers built by local Sun Action Trackers, and CPS Energy buys 100 percent of the solar power produced there—making the utility's solar capacity the largest in Texas.

More than 350 coal-fired generating units in the country have shut down during the past five years. The dilemma of aging power plants is described well in Gretchen Bakke's book *The Grid: The Fraying Wires between Americans and Our Energy Future*. Bakke reminds readers that many of the nation's plants were built for another time and have become dinosaurs in the world of power. CPS Energy has been creative in its retirement of the towering structures. The Mission Road plant, built in 1919, will be reborn in time for its hundredth birthday as the EPIcenter, a combination museum, think tank, conference center, and education hub dedicated to energy.

Award-winning architectural firm Lake Flato has produced a hypothetical rendering. Discussions continue about the best way to transform the giant shell, where generators and turbines provided almost enough electricity to run the city nearly a century ago, into a facility that preserves history and welcomes the future. The Comal plant, situated along Landa Falls in New Braunfels, has been converted into stylish lofts and shops, and reincarnations are being considered for the soon to retire Deely plant.

The Jones Avenue facility was closed in 2016, and CPS Energy conveyed a portion of the property to the San Antonio Museum of Art. The Tuttle plant in northwest San Antonio was razed, and the site has been host to the National Lineman Rodeo, in which linemen from around the country, wearing the latest in protective gear, climb poles in record time—every bit as exciting as Los Voladores at HemisFair fifty years ago.

The retired Mission Road plant will become the EPIcenter, a combination museum, think tank, conference center, and education hub dedicated to energy. Early rendering by Lake Flato Architects.

In 2016 Nora Chavez departed the CPS Energy board, and John T. Steen Jr. was appointed to represent the city's northeast quadrant. A practicing attorney who had served as the 108th secretary of state, Steen was selected to chair CPS Energy's seventy-fifth anniversary as a municipally owned utility company. Planning began for communitywide celebrations that would involve CPS Energy's entire "family," which by 2016 included 3,100 employees, 786,000 electrical customers, and 339,000 gas customers.

Through 114 substations, the utility provided energy to its customers from diverse sources—35 percent from nuclear power, 34 percent from coal, 16 percent from wind and solar renewables, and 15 percent from natural gas. Annual revenues were $2.5 billion, $336 million of which went into the city's general fund.

"The City's ownership of CPS Energy is a winning model for the community," Sculley said. She continued:

It allows for a financial return to the citizen owners, as opposed to an ROI to corporate shareholders if CPS were a private company. CPS pays the City 14 percent of its gross revenues, which has exceeded $330 million in recent years, representing a real return on investment to the citizens of San Antonio. If CPS were a private company they would pay property taxes—which would never be as high as $300 million! The revenue stream that CPS Energy provides gives the City additional financial flexibility in the management of our resources to meet the demands for city services. Without the CPS revenues, the City of San Antonio would have to impose a higher tax rate

in order to fund the current level of City services, including police protection, firefighting, and garbage collection, as well as ongoing infrastructure improvement projects that keep San Antonio in the top ten cities of the United States.

CPS Energy is proud of San Antonio's national ranking as well, and economic development and quality of life for its customers and employees have long been priorities. When it became obvious that the company had outgrown its corporate headquarters, the board began a five-year process to explore possibilities. "We started with three premises," board chair Kelley explained. "We needed to stay downtown, although there was some pressure from developers to go elsewhere. Since we serve the entire city and downtown is the center of San Antonio, that's where we need to be. Second, we wanted our project to drive economic development. And finally, it was essential to be cost-effective, achievable without a rate increase for our customers. We really have done a lot of careful thinking about all this; and I think we've developed a plan that is a 'win-win' for everyone."

The board considered expanding its headquarters by converting its parking lot into a building and connecting it to existing buildings. Recognizing how difficult it would be to maintain day-to-day operations while under construction, the idea was abandoned. The board also considered buying a tract of land downtown and building a new structure but recognized that it was the most expensive option. In 2016 the decision was made to repurpose a building on McCullough Avenue that had once served as corporate headquarters for both Valero and AT&T.

Planning is under way for a new CPS Energy headquarters that will embody energy efficiency and provide improvements to its riverfront site without rate increases for customers.

"We believe it's our best long-term option," Kelley explained, "really a thirty- to fifty-year decision; and it satisfies all of our requirements. We'll stay downtown, and we'll redevelop the river frontage, which will boost economic development. It will be cost-effective and, of course, it will be a model of energy efficiency."

Corgan, a Dallas-based architectural firm, will design the almost 500,000-square-foot headquarters, and Sundt Construction will build it.

Two towers will be joined by a three-level atrium, creating a community space for 1,200 employees who will office there. The garage will be covered with state-of-the-art photovoltaic panels that will produce electricity for the building; special insulation and glass, a water-catch for landscape irrigation, the most energy-efficient heating and cooling systems available, and cutting-edge battery storage units are just a few of the ways the headquarters will embody the future of energy when it is completed in 2020.

The future is on Mayor Ron Nirenberg's mind as well. Upon election in July 2017 he automatically became a member of the CPS Energy board. During his first days in office he oversaw passage of a resolution that called for San Antonio to join the Mayors National Climate Action Agenda in support of the Paris Agreement. The mayor, city council members, and CPS Energy CEO Gold-Williams vowed to push forward on local renewable energy programs, joining more than 300 other U.S. cities who have made the same commitment. As CPS Energy celebrates its seventy-fifth anniversary of city ownership, it is the nation's largest municipally owned gas and electric utility. There are challenges, of course, as the entire power industry foresees a rapidly changing future. When asked about some of those challenges, Kelley notes that CPS Energy has $8 billion invested in company-owned power generation and recognizes the need to carefully monitor the impact that changes in distribution systems might have on this equation. He also points to the dangers of international hacking and an increasing number of cybersecurity breaches that have led the company to focus on strong security systems to reduce its vulnerability. "Can you imagine what would happen if our lives were suddenly without power?" he asks.

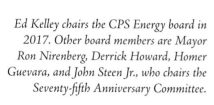

Ed Kelley chairs the CPS Energy board in 2017. Other board members are Mayor Ron Nirenberg, Derrick Howard, Homer Guevara, and John Steen Jr., who chairs the Seventy-fifth Anniversary Committee.

After storms and tornadoes in spring 2017, CPS Energy crews worked fast to restore power and repair the damage to the city. Thank you notes like these, from young children and their families, exemplify the company's "People First" motto.

Power has driven the development of the modern world, fueled by the dreams of visionary leaders and everyday citizens. Dreams remain at the heart of the CPS Energy story. They began along the banks of San Pedro Creek when the first drops of manufactured gas brought light to a dark town. As energy evolved and electricity and natural gas powered homes and businesses, the town became an international city. The population swelled, and citizens dreamed of even greater possibilities. Supplied with ample power at affordable rates, San Antonio is still growing—with more than 10,000 new houses and 3,000 new apartments projected to be built in 2017 and business start-ups on the rise.

As energy needs grow and change, as fuel sources shift and distribution systems evolve, CPS Energy will continue to adapt. "People First" is the company's enduring motto as it powers San Antonio and its dreams for the future. Following a tornado in spring 2017, several parts of the city were devastated and without power. CPS responded immediately, as it has done through countless other natural disasters, with the courtesy and kindness that have made it unique for more than a century. Praise from customers poured in.

As one young citizen and future dreamer put it, "You guys are so nice. Thank you for fixing [our] pol[e]s and other stuf[f]. . . . We are glad to have you."

APPENDIX 1

Company Names
1860–Present

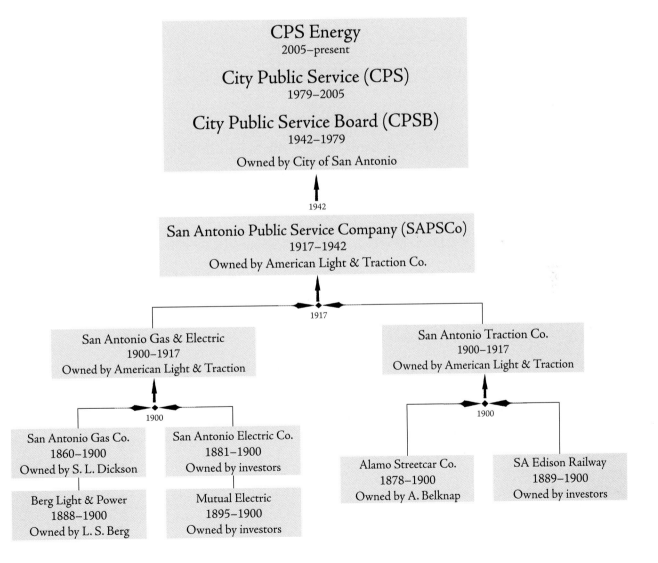

CPS Energy
2005–present

City Public Service (CPS)
1979–2005

City Public Service Board (CPSB)
1942–1979

Owned by City of San Antonio

1942

San Antonio Public Service Company (SAPSCo)
1917–1942
Owned by American Light & Traction Co.

1917

San Antonio Gas & Electric
1900–1917
Owned by American Light & Traction

San Antonio Traction Co.
1900–1917
Owned by American Light & Traction

1900

San Antonio Gas Co.
1860–1900
Owned by S. L. Dickson

San Antonio Electric Co.
1881–1900
Owned by investors

Berg Light & Power
1888–1900
Owned by L. S. Berg

Mutual Electric
1895–1900
Owned by investors

1900

Alamo Streetcar Co.
1878–1900
Owned by A. Belknap

SA Edison Railway
1889–1900
Owned by investors

APPENDIX 2

Management
1917–Present

SAN ANTONIO PUBLIC SERVICE COMPANY (SAPSCo)
1917–1942

	General Manager	President
1917–1923	Col. W. B. Tuttle	Alanson Lathrop (American Light & Traction)
1923–1933	E. F. Kifer	Col. W. B. Tuttle
1933–1936	D. A. Powell	Col. W. B. Tuttle
1936–1942	D. A. Powell	Chester Chubb (American Light & Traction)

CITY PUBLIC SERVICE BOARD (CPSB)
1942–1979

	General Manager
1942–1944	Col. W. B. Tuttle
1944–1950	E. F. Kifer
1950–1959	Victor Braunig
1959–1971	Otto H. Sommers
1971–1975	J. T. Deely
1975–1979	Jack K. Spruce

CITY PUBLIC SERVICE (CPS)

1979–2005

General Manager

1979–1985	Jack K. Spruce
1985–1999	Arthur von Rosenberg
1999–2002	Jamie A. Rochelle
2002–2005	Milton Lee

CPS ENERGY

2005–Present

General Manager/President & CEO

2005–2009	Milton Lee (General Manager)
2009–2010	Jelynne LeBlanc Burley (Acting General Manager)
2010–2015	Doyle Beneby (President & CEO)
2015–2016	Paula Gold-Williams (Interim President & CEO)
2016–Present	Paula Gold-Williams (President & CEO)

SAPSCo logo
1929–1942

City Public Service Board logo
1942–1947

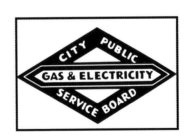

City Public Service Board logo
1947–1953

City Public Service Board logo
1953

City Public Service logo
1972–1979

City Public Service logo
1979–2005

CPS Energy logo
2005–present

APPENDIX 3

CPS Energy Trustees
1942–Present

Mayors of San Antonio
(serve as ex-officio trustees)

1942–1943	Charles K. Quin
1943–1947	Gus Mauerman
1947–1949	Alfred Callaghan
1949–1953	A. C. "Jack" White
1953*	Sam Bell Steves
1954*	R. L. Lester
1954–1955	R. N. White
1955–1961	J. Edwin Kuykendall
1961–1971	Walter W. McAllister Sr.
1971–1973	John Gatti
1973–1975	Charles L. Becker
1975–1981	Lila Cockrell
1981–1989	Henry Cisneros
1989–1991	Lila Cockrell
1991–1995	Nelson Wolff
1995–1997	William Thornton
1997–2001	Howard Peak
2001–2005	Ed Garza
2005–2009	Phil Hardberger
2009–2014	Julian Castro
2014–2017	Ivy Taylor
2017–present	Ron Nirenberg

*Mayor Jack White's terms were impacted when he suffered a series of strokes, resulting in his resignation.

CPS Energy Appointed Trustees since 1942

(in alphabetical order)

Al Aleman	1989–1994
Gen. John M. Bennett Jr.*	1950–1962
Thomas Berg*	1973–1978
Glenn Biggs*	1974–1987
Dr. Frank Bryant*	1988–1998
James H. Calvert*	1950–1962
Eloy Centeno*	1970–1980
Nora W. Chavez*	2011–2016
Lila Cockrell*	1981–1989
Leroy G. Denman Jr.*	1960–1970
Arthur Emerson*	1994–1999
Ruben Escobedo	1977–1988
Charles Foster*	2010–2011
J. H. Frost	1947–1950
Cheryl Garcia	2001–2002
John Gatti	1969–1971
Clayton Gay*	1998–2008
Charles George	1962–1965
Franz C. Groos	1942–1947
Homer Guevara*	2009–present
Stephen S. Hennigan*	2001–2008
Gloria L. Hernandez*	1989–2001
Earl C. Hill*	1978–1988
Melrose Holmgreen*	1951–1962
Derrick Howard*	2008–present
Ed Kelley*	2011–present
Pat Legan*	1987–1997
John R. Locke*	1964–1974
John H. Morse	1965–1973
Walter P. Napier*	1942–1951
John E. Newman	1971–1974
Aurora Ortega-Geis*	2002–2009
Alvaro Sanchez*	1999–2009
Willard E. Simpson*	1948–1960
John T. Steen Jr.	2016–present
Albert Steves III	1962–1969
W. B. Tuttle*	1942–1948
Dr. Robert V. West Jr.	1974–1976
Nelson Wolff*	1997–1998
D. F. Youngblood	1942–1949

*served as board chair at some point during trustee term

BIBLIOGRAPHY

PERSONAL PAPERS AND PHOTOGRAPHS

Gen. John Bennett, personal letter, John K. Matthews collection. October 1992.

Jim Berg, collection of newspaper clippings and photographs, 1970s San Antonio energy crisis.

City Council of San Antonio, official dockets. March 10, 1921, resolution to erect a memorial to Col. George Brackenridge. June 25, 1942, regular meeting of the Commissioners of the City of San Antonio.

CPS Archive, collection of historic photographs. Collection of *Broadcaster* magazines, 1922–2005; and *Lines*, 2015–2016.

Texas Transportation Museum Archive. San Antonio.

University of Texas Institute of Texan Cultures Special Collections, photograph and oral history collections.

INTERVIEWS

Berg, Tom. Civic leader and former chairman, CPSB board. Conducted by Sterlin Holmesly, December 4, 1997. University of Texas Institute of Texan Cultures Oral History Collection.

Cisneros, Henry. Former U.S. secretary of housing and urban development; former mayor of San Antonio; former city councilman. Conducted by Monica Taylor, January 27, 2017.

Cockrell, Lila. Former mayor of San Antonio; former city councilwoman. Conducted by James Dublin and Monica Taylor, September 21, 2016. Conducted by author, June 9, 2017.

Escobedo, Ruben. Former chairman, CPS Board. Conducted by James Dublin and Monica Taylor, October 26, 2016.

Fox, Lou. Former city manager, San Antonio. Conducted by author, May 22, 2017.

Fulton, Joe. Former executive, CPS Energy. Conducted by author, May, June 2017.

Gold-Williams, Paula. CEO, CPS Energy. Conducted by Monica Taylor, June 23, 2017.

Greehey Jr., William E. Chairman, NuStar Energy; former CEO/chairman, Valero Energy Corporation. Conducted by author, June 5, 2017.

Kelley, Edward. Chairman, CPS Energy Board. Conducted by author, July 5, 2017.

Sculley, Sheryl. City manager, San Antonio. Conducted by author, May 30, 2017.

Wolff, Nelson. Bexar County judge; former mayor of San Antonio. Conducted by author, May 23, 2017.

PUBLICATIONS

Allen, Paula. "Pool was part of mayor's rags to riches story." *San Antonio Express-News*, February 7, 2015.

Bakke, Gretchen. *The Grid: The Fraying Wires between Americans and Our Energy Future.* New York: Bloomsbury, 2016.

Barnhart, James A. "Quin, Charles Kennon," *Handbook of Texas Online*, accessed March 3, 2017.

Benham, Randall. "Bill Greehey." *Texas Parade* (July 1975).

Caro, Robert A. *The Path to Power: The Years of Lyndon Johnson.* New York: Alfred A. Knopf, 1982.

Coleman, Charles M. *PG&E of California: The Centennial Story of Pacific Gas & Electric Company 1852–1952.* New York: McGraw Hill, 1952.

"CPSB, Now 25 Years Old, Traces Its Origins to 1860." *San Antonio Express-News*, October 21, 1967.

Cude, Elton R. *The Wild and Free Dukedom of Bexar.* San Antonio: Munguia Printers, 1978.

"Electric History: The 1920s," *Electric Construction & Maintenance Magazine.* www.ecmweb.com.

"Feasibility of the San Antonio River Serving as a Commercial Barge Canal." *San Antonio River Authority* newsletter (summer 2012).

Fisher, Lewis. *American Venice: The Epic Story of San Antonio's River.* San Antonio: Trinity University Press, 2017.

"Gas Rate Increase Hailed." *San Antonio Express-News*, September 28, 1973.

Goho, Alexandra. "Solar Roofing Materials." *MIT Technology Review* (September 12, 2008).

Green, Mark J. "Deciding on Utilities: Public or Private?" *New York Times*, May 26, 1974.

Handy, Ryan Maye. "Utility-scale solar has grown exponentially since 2010." *MySA*, May 8, 2017.

Harder, Amy, Russell Gold, and Timothy Puko. "Trump's Support for Coal Faces First Test in Arizona." *Wall Street Journal*, February 17, 2017.

Hausman, William J. "Howard Hopson's Billion-Dollar Fraud: The Rise and Fall of American Gas and Electric." Business Conference, Frankfurt, Germany, March 13–15, 2014.

Hemphill, Hugh. *San Antonio on Wheels: The Alamo City Learns to Drive.* San Antonio: Maverick Publishing, 2009.

Henderson, Richard B. *Maury Maverick: A Political Biography.* Austin: University of Texas Press, 1970.

Hendricks, David. "Margins Helping Texas-Based New Valero Energy Corporation." *Knight-Ridder/Tribune Business News*, August 24, 1997.

"Howard Hopson, Fallen Utility Czar, Dies at 67." *Chicago Tribune*, December 23, 1949.

Hughes, L. Patrick. "Beyond Denial: Glimpses of Depression-era San Antonio." www2.austin.cc.tx.us.

Insull, Samuel. *The Memoirs of Samuel Insull: An Autobiography*, edited by Larry Plachno. New York: Transportation Trails, 1992.

Ireland, Jay. "The Key to Getting More Power to Africa." *Wall Street Journal*, March 13, 2017.

Johnson, David R., et al., editors. *The Politics of San Antonio.* Lincoln: University of Nebraska Press, 1983.

Kamp, Jon, and Kris Maher. "Coal's Decline Goes Beyond Appalachia." *Wall Street Journal*, June 20, 2017.

Lich, Glen E. *The German Texans.* San Antonio: University of Texas Institute of Texan Cultures, 1981.

"Lo-Vaca Gas Rate Approved." *San Antonio Express-News*, September 28, 1973.

Mahoney, Paul G. "The Public Utility Pyramids." Thesis, University of Virginia Law School, January 2008.

Mason, Kenneth. *African Americans and Race Relations in San Antonio, Texas, 1867–1937*. New York: Garland Publishing, Inc., 1998.

Matthews, Wilbur L. *San Antonio Lawyer: Memoranda of Cases and Clients*. San Antonio: Corona Publishing, 1983.

Mattioli, Dana, and Matt Jarzemsky. "Calpine Goes on Sales Block." *Wall Street Journal*, May 11, 2017.

"Mayor of San Antonio indicted for misapplication of funds," December 30, 1938. *Native Texans: Related Handbook of Texas Articles*.

McDonald, Forrest. *Insull: The Rise and Fall of a Billionaire Utility Tycoon*. Chicago: University of Chicago Press, 1962.

Melosi, Martin V. *Effluent America: Cities, Industry, Energy, and the Environment*. Pittsburgh: University of Pittsburgh Press, 2001.

Miller, Char. *Gifford Pinchot and the Making of Modern Environmentalism*. Washington, D.C.: Island Press, 2001.

Parson, Ellen. "Driving Improvements on the Building Energy-Efficiency Front." *Electric Construction & Maintenance*, March 22, 2017.

———. "Electrical History: The 1920s." *Electric Construction & Maintenance*, June 1, 2002.

Poole, Claire. "Stubbornness Rewarded." *Fortune*, April 13, 1992.

Puko, Tim, and Christopher Matthews. "West Texas Natural Gas Pressures Prices." *Wall Street Journal*, March 31, 2017.

Reddall, Braden, and Nichola Groom. "California Energy Storage: State Looks to 'Bottle the Sunlight' As It Moves Towards Renewable Goal." *Huffington Post*, October 12, 2013.

Rodriguez, Jason. "Moving Forward on Track: An Investigation of the Relationship between Land Use and Transportation in San Antonio." University of Texas at San Antonio, 2008.

"Roosevelt Is Glad to Be Back Again." *Daily Express*, April 7, 1905.

Russell, Jan Jarboe. "The Old Gray Mayor." *Texas Monthly* (December 2004).

Sanders, Heywood. "Empty Taps, Missing Pipes: Water Policy and Politics," in *On the Border: An Environmental History of San Antonio*, edited by Char Miller. San Antonio: Trinity University Press, 2005.

Staff of the San Antonio Express-News. *San Antonio: Our Story of 100 Years in the Alamo City*. San Antonio: Trinity University Press, 2015. Lynn Brezosky, "Prairie town to modern city"; Jennifer Hiller, "Hotels: St. Anthony known for luxury since it opened a century ago"; Neal Morton, "H-E-B: $22 billion supermarket chain started as a humble grocery store"; Ben Olivo, "Express-News building: It opened on Black Tuesday but saw brighter days"; Bill Pack, "Quarries: Retail, tourist, sports have emerged from old rock excavation sites"; Michael Quintanilla, "Joske's: Iconic Retailer once known as 'biggest store in biggest state'"; Vicki Vaughan, "Eagle Ford Shale: Field became one of the biggest developments of oil, gas"; Richard Webner, "Toyota plant: Truck maker's site on South Side viewed as economic victory."

Stacey, Daniel. "Nepal Aims to Tap Into New Energy Flows." *Wall Street Journal*, May 18, 2017.

Stevens, William K. "Texas Coal Plan: Up in Smoke?" *New York Times*, October 27, 1979.

St. John, Jeff. "Batteries Plus Backup Power: Advanced Microgrid Solutions Partners with PowerSecure." *Green Tech Media*, June 27, 2017.

Tweed, Katherine. "5 quotes We Love From BNEF's Future of Energy Summit." *Bloomberg New Energy Finance*, April 6, 2016.

Warren, Chris. "Power to the People." *San Antonio Magazine* (May 2017).

Watkins, T. H. *Righteous Pilgrim: The Life and Times of Harold L. Ickes, 1874–1953*. New York: Henry Holt, 1990.

Watson Pfeiffer, Ann Maria. "San Antonio on Track: The Suburban and Street Railway Complex through 1933." Master's thesis, Trinity University, 1982.

Wolff, Nelson. *Transforming San Antonio: An Insider's View of the AT&T Center, Toyota, the PGA Village, and the River Walk Extension.* San Antonio: Trinity University Press, 2008.

Yergin, Daniel. *The Quest: Energy, Security, and the Remaking of the Modern World.* New York: Penguin Books, 2012.

ELECTRONIC SOURCES

CPS Energy, www.cpsenergy.com

Paul W. White website, paulwhite.com

San Antonio Public Library blog, www.mysapl.wordpress.com

SAWS: History and Chronology, www.saws.org

Texas Day By Day, Texas State Historical Association, www.texasdaybyday.com

UTSA Library Digital Collections, http://digital.utsa.edu

Valero Energy Corporation, www.valero.com

Wikipedia Online Encyclopedia, www.wikipedia.org

IMAGE CREDITS

The CPS Energy Archive Photo Collection provided the majority of the photographs in this book. The author and publisher express their appreciation for this remarkable resource. Fifty additional photographs came from other sources and are credited below, again with appreciation.

Page 1. Mayor C. K. Quin, L-2756-H, UTSA Special Collections, Institute of Texan Cultures

Page 5. Sen. A. J. Wirtz, L-2744-D, UTSA Special Collections, Institute of Texan Cultures

Page 12. San Antonio de Bexar, 073-0093x8, UTSA Special Collections, Institute of Texan Cultures

Pages 14–15. Menger Hotel, 081-0480, UTSA Special Collections, Institute of Texan Cultures

Page 16. San Antonio Gas Company, 088-0320, UTSA Special Collections, Institute of Texan Cultures

Page 17. Fort Sam Houston, 108-0904a, UTSA Special Collections, Institute of Texan Cultures

Page 18. Apaches at Fort Sam Houston, 082-0654, UTSA Special Collections, Institute of Texan Cultures

Page 19. San Pedro Park, 117-0883, UTSA Special Collections, Institute of Texan Cultures

Page 21. George Brackenridge and family, L-3334-F, UTSA Special Collections, Institute of Texan Cultures

Page 23. Electricity at Post Office, 079-0377, UTSA Special Collections, Institute of Texan Cultures

Page 24. Vaudeville Theater, 101-0046, UTSA Special Collections, Institute of Texan Cultures

Page 25. Ben Thompson, Western History Collections, University of Oklahoma Libraries, Rose Collection 2146

Page 26. Electric streetcar, 086-0146, UTSA Special Collections, Institute of Texan Cultures

Page 30. Historic Pearl Brewery, courtesy of Silver Ventures and Pearl

Page 31. Sunset Station, 081-0073, UTSA Special Collections, Institute of Texan Cultures

Page 37. W. B. Tuttle, L-0313-E, UTSA Special Collections, Institute of Texan Cultures

Page 40. Early planes at Kelly Field, 069-8632, UTSA Special Collections, Institute of Texan Cultures

Page 41. Katherine Stinson, L-0851-G, UTSA Special Collections, Institute of Texan Cultures

Page 44 (top). Flood of 1921, 099-0510, UTSA Special Collections, Institute of Texan Cultures

Page 45. Flood of 1921, 080-0136, UTSA Special Collections, Institute of Texan Cultures

Page 53. Tower Life Building, UTSA Special Collections, Institute of Texan Cultures

Page 54. Unemployment lines at City Hall, L-0049-I, UTSA Special Collections, Institute of Texan Cultures

Page 55. *Time* magazine cover, courtesy of time.com

Page 56. Dizzy Dean, courtesy of Alamy, image CWAHM9

Page 59. Randolph Air Force Base, 083-0715, UTSA Special Collections, Institute of Texan Cultures

Page 61. Depression shantytown, L-1784-A, UTSA Special Collections, Institute of Texan Cultures

Page 66. Signing of the PUHCA, courtesy of Alamy, image D18FKE

Page 68. Wilson Dam, courtesy of Tennessee Valley Authority

Page 69. Buchanan Dam, L-1690-B, UTSA Special Collections, Institute of Texan Cultures

Page 71. San Antonio Municipal Airport, L-1010-J, UTSA Special Collections, Institute of Texan Cultures

Page 76. War Declared, courtesy of *San Antonio Express-News*

Page 80. Wilbur Matthews, John K. Matthews Family Collection

Page 89. San Antonio International Airport, 105-0605, UTSA Special Collections, Institute of Texan Cultures

Page 103. Hon. Lila Cockrell, E-0012-107-01, UTSA Special Collections, Institute of Texan Cultures

Page 104. Oscar and Lynn Wyatt, L-6184-B, UTSA Special Collections, Institute of Texan Cultures

Page 105. Bill Sinkin and Mayor Walter McAllister, Z-0339-62110, UTSA Special Collections, Institute of Texan Cultures

Page 106. President Kennedy, Z-1318-A-01, UTSA Special Collections, Institute of Texan Cultures

Page 111. Gov. Preston Smith, John T. Steen Jr. family collection

Page 116. Oscar Wyatt, E-0030-060-14, UTSA Special Collections, Institute of Texan Cultures

Pages 117 and 119. Energy Crisis headlines, Jim Berg Collection

Page 122. William E. Greehey, L-6715-B-14, UTSA Special Collections, Institute of Texan Cultures

Page 135. Alamodome groundbreaking, courtesy of *San Antonio Express-News*

Page 136. Visit of Pope John Paul II, courtesy of *San Antonio Express-News*

Page 141. Hon. Henry Cisneros and Hon. Ann Richards, courtesy of *San Antonio Express-News*

Page 147. *Time* magazine cover, courtesy of time.com

Page 154. AT&T Center / *Atomic Spur*, courtesy of George Cisneros

Page 156. Haven for Hope launch, courtesy of NuStar Energy

Pages 157 and 161. Mission Reach and Museum Reach, courtesy of San Antonio River Foundation

Any images appearing in this book and not on this list are from the CPS Energy Archive or are in the public domain.

ACKNOWLEDGMENTS

In early 2017 John T. Steen Jr., a CPS Energy board member and chair of the utility company's Seventy-fifth Anniversary Committee, invited me to a special meeting to explore the possibility of producing a book to commemorate the upcoming anniversary. He envisioned it as a fast-paced story that would capture the role that the nation's largest municipally owned utility company has played in transforming San Antonio. Also attending were Paula Gold-Williams, the dynamic CEO of CPS Energy; Lisa D. Lewis, CPS Energy's vice president of people and culture; and Jim Dublin, a public relations whiz and CEO of Dublin & Associates. These were the creative catalysts for the project, and I am grateful to all four of them for choosing me as their author.

Recognizing that the book would need to be completed in record time for the anniversary celebration planned for October 24, 2017, our small group approached Tom Payton, director of Trinity University Press. Despite a busy publishing schedule, he agreed that the story, which goes much back much further than seventy-five years and tracks the evolution of the nation's seventh largest city, was worth telling. We began to work at an accelerated pace to produce *Powering a City: How Energy and Big Dreams Transformed San Antonio*.

Paula Gold-Williams provided essential heart to the book project, constantly reaffirming her company's motto of "People First" and supporting my efforts to tell the story in an engaging and human way. Lisa Lewis became a thoughtful editor and adviser, contributing her profound knowledge of company history and culture. Interviews that Jim Dublin had conducted in late 2016, listed in the bibliography, provided wonderful original source material. John Steen proved to be a detail-oriented reader of the manuscript and put me in touch with some helpful contacts, including Jim Berg and John Matthews. Jim shared stories and newspaper clippings from the years that his father, Tom Berg, had chaired the board of the utility company. Those were controversial times in the energy world, and the clippings that appear in the book certainly prove it. John loaned me his personal copy of his father's book *SA Lawyer*, which describes attorney Wilbur Matthews's legal cases in detail. As the company's principal lawyer for more than five decades, Matthews Sr. worked on the sale of the utility to the City in 1942 and the lawsuit against Coastal States and Oscar Wyatt in the 1970s. Former mayor Lila Cockrell and NuStar Energy Chairman Bill Greehey also shared firsthand knowledge about the legal drama that resulted in Valero Energy's establishment in San Antonio, and NuStar's Mary Rose Brown was a human dynamo of assistance.

Interesting conversations with Bexar County judge and former San Antonio mayor Nelson Wolff, City Manager Sheryl Sculley, and former city managers Lou Fox and Terry Brechtel provided unique perspectives and helped me enrich the story in important ways. Former CPS executive Joe Fulton shared insights about the company's early sensitivity to environmental regulations, its early and continuing commitment to renewable energy, and glimpses of company culture.

When CPS Energy opened its vast archive of historic photographs and documents to me, some dating as far back as the 1860s, I was able to explore the company's evolution over a period of more than 150 years. Argel Cobb was my CPS escort to and from the archives, which are understandably under strict security regulations. I was uplifted when she told me she had read my early drafts and thought the story would appeal to a wide audience. Vincent McDonald, project manager for marketing and communications, was especially helpful as I searched through old binders for images. He produced beautiful scans for most of the 172 photographs featured in the book. Another wonderful source of historic images was the Institute of Texan Cultures Special Collections at the University of Texas, a remarkable repository that serves our city and state, directed by Amy Rushing and Tom Shelton.

The Seventy-fifth Anniversary Committee was enthusiastic and supportive of this project. Special thanks go to Monica Taylor, executive lead in the Office of the President, for helping me secure important quotes and background material, and to Zandra Pulis, deputy general counsel, for her important feedback and accuracy checks as the manuscript progressed. David L. Luschen, a senior executive at CPS Energy, gathered a group of company engineers to help me navigate some of the more technical aspects of the power business. Jill Vassar, director of development and partnerships at the EPIcenter, provided tantalizing glimpses of the transformation of the Mission Road plant, soon to be reincarnated as a museum and think tank dedicated to energy.

Every author appreciates a good copyedit, and Sarah Nawrocki provided that for this book. I am grateful to Crystal Hollis, the talented designer and production director at Seale Studios, for her beautiful work on layout, typefaces, and photographs and for meeting the accelerated production schedule with a smile. Similar thanks go to my husband, Geary Atherton, who provided me with a beautiful writing space at our little farm and tolerated my longer-than-usual hours at the computer. Finally, I have a new and immense appreciation for the gas and electricity that provide us with the heat, light, and diverse sources of power that impact our lives in such profound ways, and for the dedicated people at CPS Energy who make that possible.

INDEX

CPS ENERGY SEVENTY-FIFTH ANNIVERSARY COMMITTEE

Committee Chair, John T. Steen Jr.

Committee Advisor, James Dublin

Felecia Etheridge

KJ Feder

Yvonne Haecker

Lori Johnson

Jessica Landin

Lisa Lewis

Steffi Ockenfels

Christine Patmon

Zandra Pulis

Monica Taylor

Jonathan Tijerina

SEVENTY-FIFTH ANNIVERSARY HERITAGE SPONSORS

Special thanks to these companies for their generous support of CPS Energy through the years and for honoring CPS Energy's employees in 2017 with a gift copy of *Powering a City*.

Landis+Gyr
Norton Rose Fulbright
USAA

Catherine Nixon Cooke is the author of the biographies *Juan O'Gorman: A Confluence of Civilizations, The Thistle and the Rose: Romance, Railroads, and Big Oil in Revolutionary Mexico,* and *Tom Slick, Mystery Hunter,* which is in development as a major motion picture. She is a contributor to two anthologies, *They Lived to Tell the Tale: Adventures from the Legendary Explorers Club* and *Adventurous Dreams, Adventurous Lives.* She and her husband divide their time between San Antonio, the Texas hill country, and more remote parts of the world where untold stories beckon.